BACTERIOLOGY RESEARCH DEVELOPMENTS

MOLECULAR BASIS OF SPECIFIC MECHANISM FOR MICROBIAL ADAPTATION

BACTERIOLOGY RESEARCH DEVELOPMENTS

Additional books and e-books in this series can be found on Nova's website under the Series tab.

BACTERIOLOGY RESEARCH DEVELOPMENTS

MOLECULAR BASIS OF SPECIFIC MECHANISM FOR MICROBIAL ADAPTATION

MARCOS LÓPEZ-PÉREZ
AND
JOSE FELIX AGUIRRE GARRIDO
EDITORS

Copyright © 2020 by Nova Science Publishers, Inc.

All rights reserved. No part of this book may be reproduced, stored in a retrieval system or transmitted in any form or by any means: electronic, electrostatic, magnetic, tape, mechanical photocopying, recording or otherwise without the written permission of the Publisher.

We have partnered with Copyright Clearance Center to make it easy for you to obtain permissions to reuse content from this publication. Simply navigate to this publication's page on Nova's website and locate the "Get Permission" button below the title description. This button is linked directly to the title's permission page on copyright.com. Alternatively, you can visit copyright.com and search by title, ISBN, or ISSN.

For further questions about using the service on copyright.com, please contact:
Copyright Clearance Center
Phone: +1-(978) 750-8400   Fax: +1-(978) 750-4470   E-mail: info@copyright.com

NOTICE TO THE READER

The Publisher has taken reasonable care in the preparation of this book, but makes no expressed or implied warranty of any kind and assumes no responsibility for any errors or omissions. No liability is assumed for incidental or consequential damages in connection with or arising out of information contained in this book. The Publisher shall not be liable for any special, consequential, or exemplary damages resulting, in whole or in part, from the readers' use of, or reliance upon, this material. Any parts of this book based on government reports are so indicated and copyright is claimed for those parts to the extent applicable to compilations of such works.

Independent verification should be sought for any data, advice or recommendations contained in this book. In addition, no responsibility is assumed by the Publisher for any injury and/or damage to persons or property arising from any methods, products, instructions, ideas or otherwise contained in this publication.

This publication is designed to provide accurate and authoritative information with regard to the subject matter covered herein. It is sold with the clear understanding that the Publisher is not engaged in rendering legal or any other professional services. If legal or any other expert assistance is required, the services of a competent person should be sought. FROM A DECLARATION OF PARTICIPANTS JOINTLY ADOPTED BY A COMMITTEE OF THE AMERICAN BAR ASSOCIATION AND A COMMITTEE OF PUBLISHERS.

Additional color graphics may be available in the e-book version of this book.

Library of Congress Cataloging-in-Publication Data

ISBN: 978-1-53618-751-9

Published by Nova Science Publishers, Inc. † New York

CONTENTS

Preface vii

Chapter 1 Nonspecific Convergent Strategies
for Bacterial Adequacy to Different Types
of Environmental Stress 1
Marcos López-Pérez

Chapter 2 Molecular Mechanism Implicated in Conidia
Production by Entomopathogen Fungi 25
Divanery Rodríguez-Gómez

Chapter 3 Molecular Basis of Specific Strategies Used
by Microorganisms to Cope with Stress:
The Case of *Streptomyces* 39
*Hypatia Arano-Varela
and Francisco J. Fernández*

Chapter 4 Molecular Basis of Bacterial Homeostasis
under Environmental Stress and Cellular Transport
at Membrane Level 61
Marcos López-Pérez and Félix Aguirre Garrido

Chapter 5	Laccases, a Protein System for Adaptation through the Use of Recalcitrant Resources: Molecular Basis and Computational Modeling Approaches to Uses in the Bioindustries *E. Villegas, A. Trejo-Martínez and L. D. Herrera-Zúñiga*	**83**
Chapter 6	Host Diet and Host Derived Glycans as Primary Drivers of Microbial Gut Adaptation *Rina González-Cervantes, Félix Aguirre-Garrido and Marcos López-Pérez*	**115**
Chapter 7	Molecular Basis of Adaptation and Mechanisms Used by Halophilic Bacteria *Luis Mario Hernández Soto and José Félix Aguirre Garrido*	**131**
About the Editors		**143**
Index		**145**

PREFACE

Since the appearance of Darwin's book, "The Origin of Species," adaptation is one of the processes that explains the diversity of species in ecosystems. Adaptive phenomena in the 19th century and until the mid-20th century have been analyzed in macroscopic biological systems, however since the second half of the 20th century and to date the development of disciplines such as Molecular Biology, has allowed us to delve into the mechanisms that regulate cell physiology.

The molecular bases that allow explaining the adaptation processes of microorganisms to their environment have special relevance, because through their analysis it is possible to size the complexity of these mechanisms that involve receptors of a protein nature associated with transduction chains that transport the information flow to genomic DNA, and which subsequently involves the emission of a response through the expression of specific genes.

From the point of view of the adaptive phenomenon analysis, the approach through the molecular bases makes it possible to understand the enormous diversity of the microbial world. Mainly for two reasons, on the one hand the presence of micro gradients in the bacterial ecological niches that are continuously fluctuating, which forces the microorganisms to a rapid adaptation phenomenon. And on the other hand, horizontal gene transfer phenomena, which allow bacteria the information exchange. These two elements carry great intensity in establishing new relationships. This phenomenon is especially relevant if it is related to a concept that Darwin cites in the Origin of Species, "The tangled riverbank", where it is emphasized that the new interactions establishment is the basic driving force for the new species generation. This mechanism is explained by the positive feedback loop generation, whereby ecosystems with high levels of biological diversity generate new interactions that lead to new species, which in turn tends to make the ecosystem network more complex. This complexity analyzed in its molecular bases allows to generate new research questions that can be applied to other knowledge areas, such as Biotechnology.

The analysis of the molecular bases of the microorganisms adequacy, makes it possible to identify and characterize mechanisms that implemented in different pharmaceutical areas and agricultural industry has led to the product generation with high added value, a clear example of this economic development is the enzyme industry and even recombinant protein production.

Finally, it is convenient to emphasize the need to incorporate the analysis of the molecular bases of adaptation from the perspective of omics techniques. Techniques that allow the study of processes and mechanisms to be approached from a global perspective.

This book summarizes some topics of special relevance referring to adaptive processes of different microorganisms of special relevance both in basic and applied research.

In: Molecular Basis of Specific Mechanism ... ISBN: 978-1-53618-751-9
Editors: Marcos López-Pérez et al. © 2020 Nova Science Publishers, Inc.

Chapter 1

NONSPECIFIC CONVERGENT STRATEGIES FOR BACTERIAL ADEQUACY TO DIFFERENT TYPES OF ENVIRONMENTAL STRESS

Marcos López-Pérez[*]
Environmental Sciences Department,
Metropolitan Autonomous University (Lerma Unit)
Lerma de Villada, México

ABSTRACT

Bacteria and, in general, less complex organisms, from the point of view of functioning as a biological system, have been the most successful in the tree of life. On the one hand, they have reached a complete distribution in all ecological niches of the biosphere and, on the other, their success is confirmed by their permanence from the origin of life on Earth until today. These two elements, placed into the perspective of natural history, show that bacteria have an enormous adaptive capacity. The purpose of this chapter is to review and deepen the nonspecific bacterial systems and mechanisms that allow for the adaptation of bacteria to eventual changes in the variables (biotic and abiotic) of ecosystems.

Keywords: nonspecific stretegies, bacterial, stress

---

[*] Corresponding Author's Email: m.lopez@correo.ler.uam.mx.

1. Introduction

The biosphere's history can be understood as the history of a constant change, the intensity of this change has also fluctuated, being higher in the period during which life could have emerged on the planet (Grant et al., 2017). The selection pressure in ecosystems can be defined as the variability of the different types of stress present in ecosystems (biotic and abiotic). This selection pressure leads over time to evolutionary changes (Hoffmann et al., 2000), consequently, to speciation processes that come with the emergence of new genotypes expressing a variety of phenotypes. The processes that enable this dynamics are the adaptation mechanisms, among which we can distinguish two categories: a) non-specific mechanisms, meaning those processes executed in a non-specific way to mitigate or limit the effects generated by a particular environmental context that can be identified as stressful. This type of mechanism is characterized by being fast and less efficient in its objective if compared to the other category; b) specific mechanisms, which, in general, have a longer delay in expressing themselves, but are more efficient in terms of energy cost (Tyukina et al., 2016) to mitigate, limit, or suppress the effects derived from the environmental stress. On the other hand, it is very important to elucidate the operation dynamics of these mechanisms to characterize the ecosystems at the microbial scale, as well as to contextualize the phenomena of cellular response to environmental changes under these conditions. Two fundamental parameters to figure out the adaptive phenomena in bacteria should be considered; on one side, the relevance of population densities must be reviewed and, on the other, the diversity of metabolites to which a bacterial cell could potentially be exposed. This chapter aims to review those mechanisms that can be categorized as nonspecific and that are common response pathways to environmental changes of a very different nature (biotic and abiotic), such as the presence of antibiotics, heavy metals, environmental pollutants or high salt concentrations, presence of metabolites of other organisms, phenomena related to the transport and maintenance of cell homeostasis. Finally, an element that must be considered is that the operation of nonspecific mechanisms facilitates a rapid response to an environmental change that aims to limit or mitigate a physiological effect derived from a stressful situation, however, under this response context, they still exert an effect. Specific adaptation mechanisms are the ones to have a greater efficiency in eliminating the effect derived from this stressful environmental situation.

2. Concept of Adaptive Landscape and Optimal Evolutions in Microorganisms

The coevolution process analysis in microbial ecology is especially important to elucidate the stability of bacterial nonspecific adaptation mechanisms; for this process, it is particularly relevant to take into account that an adaptation mechanism can be characterized by the following assumptions: a) it remains within the genetic background during long periods of time, as can be inferred from relevant studies (Thompson et al., 2002), b) it has extended to other organisms (Li & Nikaido 2009), and c) it is derived from coevolution processes, particularly among microbial species that share the same ecological niche (Lammers & Freeman 1986; Apaloo et al., 2009). To understand better the coevolution process, it may be necessary to introduce the fitness landscape notion, originally coined in 1939 (Wright, 1939) and studied later (Kauffman & Levin, 1987), which has been taking shape in recent years (de Visser et al., 2018). Thus, in the evolutionary biology conceptual framework, fitness landscapes are used basically to establish a relationship between two parameters: the genotype and reproductive success. When a particular character has as a consequence that this relationship has a maximum value, then it is possible to infer that this feature is optimized. This concept has been widely described for the size and volume of bacteria. For example, Koch observed that the lower boundary of prokaryotic cell size is the one "large enough to house the total amount of needed stuff" (Koch, 1996). In addition, these studies are reinforced by the theory that the speed of chemical reactions and their effect on the cell, particularly protein-substrate and protein-

protein interactions, are conditioned by the confinement imposed by the membrane to the cell metabolism (Küchler et al., 2016). From these studies, it is possible to infer that bacterial size and shape are optimized characteristics and any deviation from this parameter will be penalized by natural selection. This example makes it possible to understand the nonspecific mechanisms to adaptation, particularly the cellular efflux systems. These systems can be considered as processes that are in an optimal evolutionary point in the adaptive landscape displayed at the initial stage in the cellular response to environmental changes. Their function is therefore to initially mitigate an environmental disturbance in a quick and non-specific way. Other elements supporting the optimization of cellular efflux systems are their relation to other processes also used for the adequacy, as for example the presence of heavy metals (Chen et al., 2015), the presence of bacteriophages (Haaber et al., 2016), or the temperature increase (Bengoechea and Skurnik 2000). On the other hand, it is possible to infer that when a feature is optimized, it inexorably leads to the development of co-evolutive processes from other organisms that share their ecological niche (Kauffman and Johnsen; 1991Carneiro and Hartl, 2010). In this regard, antibiotic therapies cannot be considered optimized as in almost 100 years they have not achieved sufficient stability. As a consequence, multiple resistances were generated and new resistances are still appearing (Barlow, 2018). In this sense, it is pertinent to cite works that deepen the processes of adaptation to the presence of antibiotics through coevolution mechanisms (Bottery el al., 2017). Another important process of coevolution is that observed during the pathogenesis relationship occurring between bacteria and bacteriophages. For example, recently, new mechanisms of evolution, developed by a lineage of phage λ, have been described (Burmeister et al., 2016). Thus, bacteriophages, unlike what happens with antibiotics, have been able to circumvent bacterial resistance mechanisms for thousands of millions of years through the development of co-evolutive dynamics. This is eventually based on the change of interaction patterns among the different elements found in bacteria and bacteriophages.

3. THE MICROBIAL SCALE ADAPTIVE LANDSCAPE AND ITS IMPLICATIONS ON ITS OPERATING DYNAMICS

The adaptation mechanisms developed by organisms throughout their natural history must be understood and contextualized according to the ecosystem characteristics of the niche where they live. In this sense, it is necessary to characterize the conditions of the ecosystems where microorganisms live, develop, adapt, and evolve. For this reason, it is interesting to highlight one of the variables that determines the microorganism's physiology, their small size, and, therefore, their enormous area/volume ratio. This condition that is common to all microorganisms (bacteria, fungi, yeasts, and even viruses) has a very relevant consequence, which has been published in important works and can be summarized as that microorganisms are very much exposed to environmental conditions (Harris and Theriot 2018). This characteristic evolved in a convergent manner in all microorganisms, as well as other characteristics of great relevance in bacterial physiology, as published in relevant works (Stern, 2013). On the other hand, it is necessary to deepen the knowledge of ecological niches characteristics at a microscopic scale; in this sense, niches of aquatic ecosystems and niches in terrestrial ecosystems must be distinguished. For niches in aquatic ecosystems, the presence of water in large volumes has as a direct consequence, i.e., the dissolution of many of the elements that could eventually affect microbial physiology, resulting in an attenuation of a possible derivative effect. On the other hand, the physicochemical characteristics of water facilitate homeostatic regulation since the abiotic parameters of these ecosystems tend to be confined in ranges of values that are very stable over time (Torday, 2015). Therefore, in these ecosystems, it can be said that the presence of gradients should be considered in enormous volumes and distances with respect to the average size of the microorganisms present in these niches. It is also necessary to indicate that the displacement of these organisms is much higher on average than that of terrestrial ecosystems microorganisms as it depends directly on external agents that induce the movement of water due to changes in temperature, salinity, and density

(Cabral, 2010). This huge displacement with respect to their size results in exposure to different environmental conditions, which have necessarily conditioned the phenomena of adaptive response in bacteria that inhabit these ecological niches. On the other hand, for terrestrial ecosystems, it is necessary to emphasize the ecosystem that predominantly harbors mosst bacteria, the soil (Zhang et al., 2013). The soil is characterized by being a heterogeneous niche, where variables such as texture, chemical composition, or pH condition the water retention capacity, water penetration capacity, presence of oxygen, or reactions of a very diverse nature with microbial metabolism products (Rousk, 2011). Precisely this heterogeneity results in the presence of a very different nature of gradients in a small volumes or spaces in the soil respect to the average size of the bacteria. Although, in the terrestrial environment, the movement of microorganisms is limited because it depends on the presence of external agents such as water or wind, this heterogeneity has the same effect as the large displacement of bacteria present in aquatic ecosystems, that is, the need to develop systems and mechanisms that facilitate the rapid adaptation to very fluctuating environmental conditions and, consequently, causing stress. These elements have determined the bacteria's convergent evolution, in the sense of developing non-specific adaptation mechanisms, understanding that this is the first mechanism that is started in an eventual adaptation process and, once the type of stress or environmental variable has been characterized in bacterial physiology, the response is adjusted by becoming more specific and, therefore, more energy-efficient.

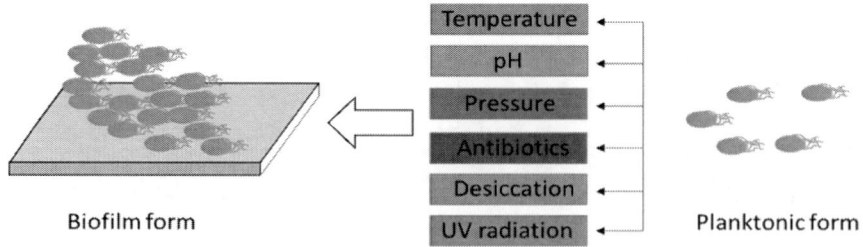

Figure 1. Environmental variables that can act as elicitors of cell aggregation for the formation of biofilms.

4. RELEVANCE OF THE BACTERIAL ORGANIZATION IN ADAPTIVE PHENOMENA

One of the main nonspecific mechanisms that by evolutionary convergence phenomena spread throughout the microbial world is related to the ability to associate by forming biofilms (Jefferson, 2004). In this sense, it is convenient to emphasize that microbial communities under certain environmental conditions are capable of self-aggregation by additionally producing an extracellular matrix consisting mainly of polymeric substances (EPS), constituting a structure of recalcitrant nature, which confers protection to bacteria from ultraviolet radiation, high temperatures, extreme pH values, high pressures, low concentrations of nutrients, antibiotics, among others (Yin et al., 2019) (Figure 1). This aggregation allows microbial communities to attenuate the effect of environmental factors, since this phenomenon enables the generation of microgradients.

In this sense, this effect has already been published in situations of growth inhibition due to high concentrations of substrate or catabolic repression (Viniegra & Favela, 2005), where it has been determined that cell aggregation makes it possible to gradually decrease the concentration of the exterior of the biofilm with respect to the interior of the aggregate, a gradient that allows adaptation through the progressive mitigation of a stressful situation, probably because the exopolysaccharide matrix and cellular compartmentalization facilitate evading the potential direct effect on cells. On the other hand, this phenomenon has been studied regarding the presence of antibiotics, determining that the presence of antibiotics induces cell aggregation by activating the quorum sensing system (Jiang et al., 2019). Although the presence of antibiotics in pristine environments has not been determined, it is intuited that they are present in trace amounts with respect to the concentrations found in hospital centers and patients under antibiotic therapy, which is why aggregation can be understood as an unspecific measure. In the first place, because it generates gradients for any potential agent of the medium causing stress, and secondly because this response has been determined and characterized throughout the realm of bacteria, and even for single-celled eukaryotes, such as yeasts (Lima Pérez et al., 2017).

5. Presence of Secondary Metabolites

In the microbial world, the most relevant element to consider from a generic perspective refers to the forced interaction with other microorganisms forming physiological groups and communities. In this sense, the interaction phenomena are directly related to the presence of secondary metabolites released by microorganisms seeking to elicit an effect on other organisms. This cellular communication network through secondary messengers allows microorganisms to execute a coordinated response to environmental conditions. Although the variety of secondary metabolites is huge, in this area it is convenient to focus on antibiotics because of their greater presence in the literature. In this sense, it has already been argued in different relevant works that the so-called "resistance mechanisms" can be understood, in the microbial ecology of bacteria, as ways of collecting and transferring information from one cell to another (Fajardo et al., 2009), however, derived from the anthropocentric vision promoted by clinical microbiology, antibiotics have always been considered as compounds with bactericidal or bacteriostatic action, leaving aside their role as regulators of the Ecophysiology of physiological groups and bacterial communities.

6. Nonspecific Antibiotics Resistance Mechanisms

The mechanisms of antibiotics resistance can be categorized into two groups: specific, mainly when it comes to the direct enzymatic action on antibiotics, and non-specific, when they try to mitigate the effect of these secondary metabolites, but only partially achieve it. The mechanisms and molecular bases described in this area have been found in a wide range of bacteria. It is interesting to highlight that both antibiotics and the mechanisms of adaptation or resistance to them appeared at a previous time in natural history, so we can say that they are widely extended in time (Waglechner & Wright, 2016). The resistance phenomenon to antimicrobial drugs with bactericidal or bacteriostatic effect in therapies for the bacterial suppression in sepsis has been described since the beginning of the antibiotic era. Currently, there are many works that describe the mechanisms through which bacteria are able to circumvent the effect of antibiotics (Blair et al., 2015; Munita & Arias., 2016). In addition, it is pertinent to clarify that the understanding of mechanisms that confer resistance is essential for the further identification of genes and, consequently, of proteins involved in these processes, because the enzyme action can be inferred from the intrinsic characteristic of the resistance mechanism and, therefore, the search for genetic sequences or specific enzyme activities through different techniques can be implemented. Below, are the main mechanisms by which bacteria evade or attenuate the action of these compounds. These mechanisms are depicted on Figure 2.

6.1. Modifications in the Target Molecule

This mechanism is based on the structural modification in the target molecule that interacts with the antibiotic. In addition, and in terms of the cellular physiology of the microorganism carrier, it is pertinent to clarify that this molecular change does not have a significant effect on the target molecule catalysis, enabling a catalytic action with the same or similar efficiency that the original molecule, avoiding therefore the specific binding to the antibiotic. A relevant example refers the modification of the penicillin-binding proteins (PBPs) in *Streptococcus pneumoniae* (Tsurusako et al., 2009), generating resistance to penicillin. Another similar strategy within the general framework of this mechanism is to produce a ligand very similar structurally to the target molecule, whose function is to bind the antibiotic. An example of this mechanism is the PBP2 protein described in *Staphylococcus aureus* (Barrett et al., 2005). This mechanism significantly reduces the number of interactions between the drugs and the target molecule. This type of resistance has been reported for antibiotics at any step in the process of protein synthesis (Poehlsgaard & Douthwaite, 2005).

6.2. Permeability Alterations

This mechanism includes several types of strategies for the generation of resistance, whose common element is that this resistance is not total or can be understood as a partial resistance. A) The first strategy has been studied in gram-negative bacteria, where the membranes prevent the transport of hydrophilic substances, which has as a consequence that the transport must occur through proteins with porin function. The mechanism is based on the downregulated transcription of genes encoding these proteins, as has been described for antibiotic resistance in *Pseudomonas aeruginosa* (Ochs et al., 1999). B) Another specific case that is included in this general mechanism refers to modifications or alterations in the antibiotics penetration mediated by proteins that require energy. This mechanism has been described for aminoglycosides, where it has been determined that the entry involves adherence to molecules with a net negative charge, such as lipopolysaccharide (LPS) or the heads of membrane phospholipids that once stick access the periplasmic space and, finally, enter the cytoplasm by a system coupled to a proton gradient. The resistance mechanism associated with this system is related to the proton gradient modification, which consequently difficult access of these molecules to the cytoplasm (Moffatt et al., 2010). C) Finally, the latter mechanism of this group refers to the antibiotics expulsion, also known as resistance due or caused by efflux pumps. As previously mentioned, it is a nonspecific mechanism, which has been studied for beta-lactam antibiotics, quinolones, tetracyclines, and chloramphenicol. In particular, it has been determined that, in gram-negative bacteria, the mechanism requires three proteins: one associated to the cytoplasmic membrane, another protein whose function has been associated with the fusion of the two membranes, and a protein associated with the outer membrane and porin function, the most described are Mex AB-Pro M, Mex CD-Pro J, and Mex EF-OprN (Poole, 2000).

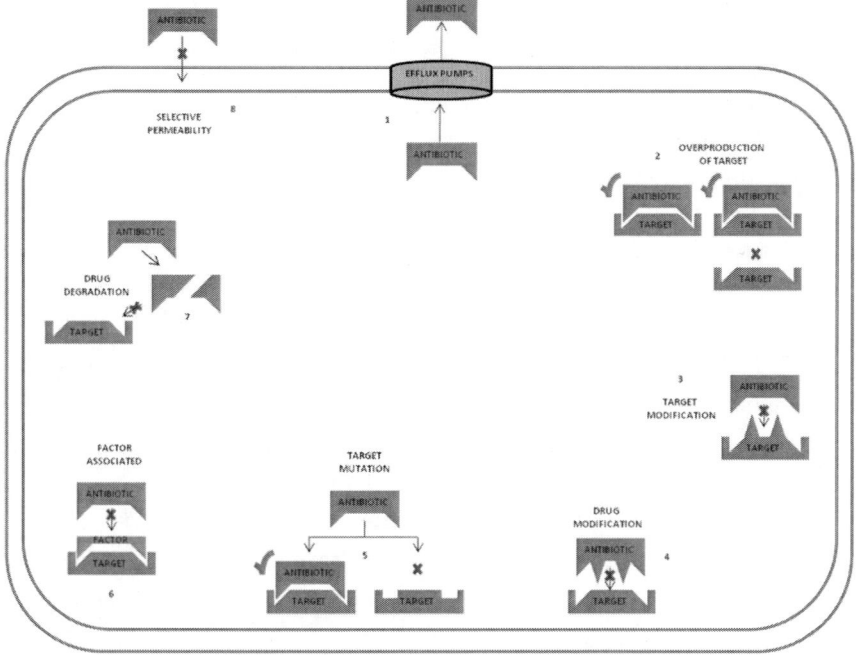

Figure 2. General mechanism for antibiotics resistance.

6.3. Overexpression of Target Molecules

In this mechanism, the strategy to generate resistance is based on the overproduction of the molecule that binds the antibiotic, in this way only part of the population of the target molecules become inactivated; thus, cellular functions are not completely inhibited. A relevant example of this mechanism is based on the overexpression of a fragment of rRNA similar to h34 of the 16S rRNA gene, which confers resistance to spectinomycin (Thom & Prescott, 1997).

6.4. Drug Modification and Degradation

In the literature, there are many description of proteins whose catalytic action is to degrade antibiotics (Wright, 2005; Ramírez & Tolmasky 2010). The enzyme catalysis, described in this mechanism of resistance, involves adenylation, acetylation, phosphorylation, glycosylation (Alekshun and Levy, 2007). On the other hand, it is interesting to note that the enzyme specificity has a high efficiency in the inactivation of the antibiotic, which implies that this type of mechanism confers total resistance to the presence of an antibiotic.

7. Cellular Transportation

The fundamental objective of any process or mechanism focused on rapid adaptation in the microbial world is based on maintaining homeostatic balance, a balance that allows the values of the variables that define the internal cellular environment to be maintained in a reasonably stable interval. In this sense, it is pertinent to refer to the main cellular function that quickly allows bacteria to adapt and transport. In general, it is possible to relate the phenomenon of transport with the attenuation of the potential effect caused by biotic and abiotic variables in the bacterial context. Next, we will go deeper into the transport area most published in the literature, mainly the presence of heavy metals and the regulation of pH, which is finally a parameter that globally conditions transport in cellular physiology.

8. Transportation of Metals Associated with Adequacy

One of the processes that allows the homeostatic regulation of the cell is transport, as has been published repeatedly (Dubyak, 2004). Several works have dealt with heavy metal transporters like magnesium, which is a fundamental element for ribosomes, stabilization of membranes, as well as for the neutralization of nucleic acids and as a cofactor in a variety of enzymatic reactions (Groisman et al., 2013). It has been described that cells have developed a mechanism of several types of transporters (Hmiel et al., 1989), as well as their mechanism for regulating magnesium transporters (Miller et al., 2001). This mechanism can be considered as nonspecific because it affects basic functions of bacterial metabolism and, additionally, it is widely represented in the bacterial phylogenetic tree (Maguire, 2006). Another clear example are the transporters for copper, an essential micronutrient as a cofactor for different enzymes (Argüello et al., 2013). Regulation of copper transport should be understood as a nonspecific adaptation mechanism because it is directly related to the generation of chemical energy, particularly through the family of respiratory oxidases or cuproenzymes (Garcia-Horsman et al., 1994; Richter and Ludwig, 2003), responsible for the reduction of oxygen and the consequent generation of the electrochemical proton gradient (Garcia-Horsman et al., 1994). Multiple mechanisms that relate the transport of metals to cellular homeostasis have been described in relevant works (Chandrangsu et al., 2017). Considering the previously mentioned works, it is possible to conclude that the regulation of this transport can be understood as non-specific because it affects the main element used by bacteria to achieve adaptation, i.e., homeostatic regulation.

9. MOLECULAR ASPECTS OF BACTERIAL pH SENSING FOR ADAPTATION

The molecular bases of multiple mechanisms that allow bacteria to tolerate, grow and, consequently, adapt to environmental pH conditions are described in the literature. Particularly interesting are the mechanisms described for extremophile organisms, capable of adapting to a pH range of approximately 1 to 13. One of the nonspecific mechanisms of great relevance refers to the regulation of the constitutive expression of proteins related to the attenuation of the impact generated by sudden changes in pH. It has been reported that the presence of these mechanisms in bacteria under neutral pH conditions leads to a high energy cost due to the fact that they are constitutive and because many of the proteins are adapted to work under extreme pH conditions, functioning less efficiently under neutral pH (Gilmour et al., 2000; Hicks et al. 2010). In general terms, the main strategy for homeostasis control in relation to the pH refers to the synthesis of active proton transporters (Padan et al., 2005). Under acidic conditions, for neutroalophile bacteria, the mechanism reacts by increasing the expression levels of the respiratory chain complexes that pump protons out of the cell and, at the same time, couple to the ATSase complex, assist in synthesizing ATP, and derive more energy flow to cellular maintenance (Kobayashi et al., 1986) (Figure 3). However, for alkaline pH, the most widespread nonspecific mechanism is the overexpression of active proton transporters to the cytoplasm (Maurer et al., 2005) (Figure 3).

In mesophilic microorganisms, the homeostasis maintenance in relation to the pH is based on increased enzymes expression related with the transport of protons from the cytoplasm to outside the cell, among these enzymes are hydrogenases and amino acid decarboxylases (Maurer et al., 2005; Stancik et al., 2002). In an acid environment, *E. coli* expresses the hydrogenated systems that catalyze the production of H_2 from protons in the cytoplasm, contributing to raising the pH, which consequently enables survival at ambient pH values between 2 and 2.5 (Noguchi et al., 2010).

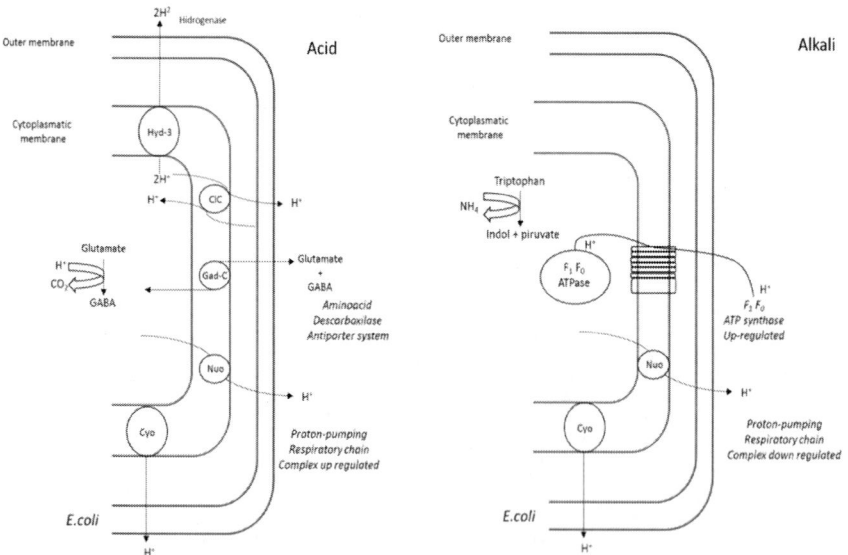

Figure 3. Homeostatic maintenance strategy by regulating the consumption of protons and the production of metabolic enzymes.

10. ENVELOPE STRESS RESPONSES (ESRS)

One of the nonspecific mechanisms that is shared by a huge variety of organisms with adaptive effects refers to the organization, structure, and composition of membranes (Shrivastava and Chang, 2019). Envelope biogenesis is a complex process that involves several machineries (Silhavy et al., 2010). Falta algo. Surprisingly, some bacteria are able to get rid of their wall under the influence of stress, producing cells that are deficient in the cell wall. In particular, wall-deficient cells are flexible and can maneuver through narrow spaces, insensitive to antibiotics directed to the wall and capable of absorbing and exchanging DNA. In addition, since epitopes associated with the wall are often recognized by host defense systems, the wall deficiency provides a plausible explanation of how some bacteria can hide in their host. In this review, we focus on this paradoxical stress response, which gives cells unique opportunities that are not available for walled cells. (Claessen, 2019). For extremophile organisms, it should be noted that although the adaptive response is much more efficient, it is not appropriate to characterize these mechanisms as nonspecific, since the ecosystems of extremophile organisms are characterized by being stable over time. Some extreme acidophiles are heterotrophic bacteria, including the thermo-acidophilic *Alicyclobacillus acidocaldarius*, which contains an unusual ω-alicyclic fatty acid as a major membrane component (Wisotzkey, 1992). On the other hand, recently published works have detailed the relevance of the size of the periplasmic space (controlling the inner membrane-to-outer membrane distance), particularly in gram-negative bacteria, as a fundamental element that regulates the communication phenomenon, finding that this size is highly regulated to facilitate or hinder certain signals that must be perceived inside the cell (Asmar et al., 2017). It has been argued that evolution through convergence processes has led these bacteria to optimize this variable as an element of nonspecific response to environmental stress. Finally, one of the variables that the cells, in relation to their envelopes, have developed as an element of response to environmental change refers to cell volume. It is pertinent to highlight that cell volume is the baseline parameter that defines the molar concentrations intracellularly, therefore, it is the element that conditions the interaction phenomena among the different cellular components, like enzyme-substrate interaction, detection of cellular second messengers (cAMP), or the interaction phenomena with DNA transcription factors, among others. In this sense, it has been published that, surprisingly, the enzymes inside the cell are limited by the presence of substrate, having measured less enzymatic activity inside the cell compared to *in vitro* analyses: the production of enzymes being more energy-efficient than the uptake of substrate (Zotter et al., 2017).

CONCLUSION

The microbial world is characterized by the great parameters variability present in the ecosystem that additionally fluctuate rapidly. This ecological niche, has forced microorganisms to develop rapid adaptation mechanisms to ensure survival. Non-specific adaptation mechanisms allow microorganisms to mitigate the unforeseen effect of changing an environmental variable. Mechanisms that are fundamentally focused on the cellular homeostasis maintenance and therefore have implications for all cellular physiology.

REFERENCES

Alekshun, M. & Levy, S. (2007). Molecular mechanisms of antibacterial multidrug resistance. *Cell*, *12*, 1037-1050. https://doi.org/10.1016/j.cell.2007.03.004.

Apaloo, J., Brown, J. S. & Vincent, T. L. (2009). Evolutionary game theory: ESS, convergence stability, & NIS. *Evol Ecol Res*, *11*, 489–515. https://doi.org/10.1007/978-3-319-44374-4_6.

Argüello, J. M., Raimunda, D. & Padilla-Benavides, T. (2013). Mechanisms of copper homeostasis in bacteria. *Frontiers in cellular and infection Microbiology*, *3*, 73. https://dx.doi.org/10.3389%2Ffcimb.2013.00073.

Asmar, A. T., Ferreira, J. L., Cohen, E. J., Cho, S. H., Beeby, M., Hughes, K. T. & Collet, J. F. (2017). Communication across the bacterial cell envelope depends on the size of the periplasm. *PLoS Biol.*, *19*, 15(12):e2004303. doi: 10.1371/journal.pbio.2004303.

Barlow, G. (2018). Clinical challenges in antimicrobial resistance. *Nat Microbiol*, *3*, 258–260. https://doi.org/10.1038/s41564-018-0121-y.

Barrett, D., Leimkuhler, C., Chen, L., Walker, D., Kahne, D. & Walker S. (2005). Kinetic Characterization of the Glycosyltransferase Module of Staphylococcus aureus PBP2. *Journal of Bacteriology*, (187) 6, 2215–2217. https://dx.doi.org/10.1128%2FJB.187.6.2215-2217.2005.

Bengoechea, J. A. & Skurnik, M. (2000). Temperature-regulated efflux pump/potassium antiporter system mediates resistance to cationic antimicrobial peptides in Yersinia. *Molecular Microbiology.*, *37*, 67-80. https://doi.org/10.1046/j.1365-2958.2000.01956.x.

Blair, J., Webber, M., Baylay, A., Ogbolu, D. & Piddock, L. (2014). Molecular mechanisms of antibiotic resistance. *Nature reviews Microbiology.*, 1310-1038. doi: 10.1038/nrmicro3380.

Bottery, M. J., Wood, A. J. & Brockhurst, M. A. (2019). Temporal dynamics of bacteria-plasmid coevolution under antibiotic selection. *ISME J*, *13*, 559–562. https://doi.org/10.1038/s41396-018-0276-9.

Burmeister, A. R., Lenski, R. E. & Meyer, J. R. (2016). Host coevolution alters the adaptive landscape of a virus. *Proc Biol Sci.*, (28), 283(1839). https://dx.doi.org/10.1098%2Frspb.2016.1528.

Cabral, J. P. (2010). Water microbiology. Bacterial pathogens and water. *International journal of environmental research and public health*, *7*(10), 3657–3703. https://dx.doi.org/10.3390%2Fijerph7103657.

Carneiro, M. & Hartl, D. L. (2010). Adaptive landscapes and protein evolution *Proceedings of the National Academy of Sciences*, *107*, 1747-1751. doi: 10.2307/40536021.

Chandrangsu, P., Rensing, C. & Helmann, J. D. (2017). Metal homeostasis and resistance in bacteria. *Nature reviews Microbiology*, *15*(6), 338–350. https://doi.org/10.1038/nrmicro.2017.15.

Chen, H., Sijin, Y. T., Wang, L. Y. & Wang, J. (2015). *Contamination features and health risk of soil heavy metals in China Science of The*

Total Environment, 143-153. https://doi.org/10.1016/j.scitotenv.2015. 01.025.

Claessen, D. & Errington, J. (2019). Cell Wall Deficiency as a Coping Strategy for Stress *Trends in Microbiology*, *27* (12), 1025-1033, https://doi.org/10.1016/j.tim.2019.07.008.

de Visser, J. A., Elena, S., Fragata, I. & Matuszewski, S. (2018). The utility of fitness landscapes and big data for predicting evolution. *Heredity*, *121*, 401–405. https://doi.org/10.1038/s41437-018-0128-4.

Dubyak, G. R. (2004). Ion homeostasis, channels, and transporters: an update on cellular mechanisms *Advances in Physiology Education*, *28*, 4, 143-154. https://doi.org/10.1152/advan.00046.2004.

Fajardo, A., Linares, J. F. & Martínez, J. L. (2009). Towards an ecological approach to antibiotics and antibiotic resistance genes. *Clin. Microbiol. Infect.*, *1*, 14–16. https://doi.org/10.1111/j.1469-0691.2008.02688.x.

Garcia-Horsman, J. A., Barquera, B., Rumbley, J., Ma, J. & Gennis, R. B. (1994). The superfamily of heme-copper respiratory oxidases. *J. Bacteriol.*, *176*, 5587–5600, 355. https://dx.doi.org/10.1128%2Fjb.176. 18.5587-5600.1994.

Gilmour, R., Messner, P., Guffanti, A., Kent, R., Scheberl, A., Kendrick, N. & Krulwich, T. (2000). Two- Dimensional Gel Electrophoresis Analyses of pH-Dependent Protein Expression in Facultatively Alkaliphilic *Bacillus pseudofirmus* OF4 Lead to Characterization of an S-Layer Protein with a Role in Alkaliphily. *Journal of bacteriology.*, *182*, 5969-81. https://dx.doi.org/10.1128%2Fjb.182.21.5969-5981. 2000.

Grant, P. R., Grant, B. R., Huey, R. B., Johnson, M. T. J., Knoll, A. H. & Schmitt, J. (2017). Evolution caused by extreme events. *Philos Trans R Soc Lond B Biol Sci.*, *19*, 372(1723), 20160146. doi: 10.1098/ rstb.2016.0146.

Groisman, E. A., Hollands, K., Kriner, M. A., Lee, E. J., Park, S. Y. & Pontes, M. H. (2013). Bacterial Mg2+ homeostasis, transport, and virulence. *Annual review of genetics*, *47*, 625–646. doi: 10.1146/ annurev-genet-051313-051025.

Haaber, J., Leisner, J., Cohn, M., Catalan-Moreno, A., Nielsen, J., Westh, H., Penadés, J. & Ingmer, H. (2016). Bacterial viruses enable their host to acquire antibiotic resistance genes from neighbouring cells. *Nature Communications.*, *7*, 10.1038. https://doi.org/10.1038/ncomms13333.

Harris, L. K. & Theriot, J. A. (2018). Surface Area to Volume Ratio: A Natural Variable for Bacterial Morphogenesis. *Trends Microbiol.*, *26*(10), 815-832. https://doi.org/10.1016/j.tim.2018.04.008.

Hicks, D. B., Liu, J., Fujisawa, M. & Krulwich, T. A. (2010). F1F0-ATP synthases ofalkaliphilic bacteria: lessons from their adaptations. *Biochim. Biophys. Acta.*, *1797*, 1362–1377. https://dx.doi.org/10.1016%2Fj.bbabio.2010.02.028.

Hmiel, S. P., Snavely, M. D., Florer, J. B., Maguire, M. E. & Miller, C. G. (1989). Magnesium transport in *Salmonella typhimurium*: genetic characterization and cloning of three magnesium transport loci. *Journal of bacteriology.*, *171*(9), 4742–4751. https://dx.doi.org/10.1128%2Fjb.171.9.4742-4751.1989.

Hoffmann, A. A. & Hercus, M. J. (2000). Environmental Stress as an Evolutionary Force, *BioScience*, *50*, (3), 217–226. https://doi.org/10.1641/0006-3568(2000)050[0217:ESAAEF]2.3.CO;2.

Ingledew, W. J. (1990). *Microbiology of Extreme Environments*. Edwards, C., editor. McGraw-Hill, 33-54. ISBN-10: 0335158935.

Jefferson, K. K. (2004). What drives bacteria to produce a biofilm? *FEMS Microbiology Letters*, *236*, (2) 163–173. doi: 10.1016/j.femsle.2004.06.005.

Kauffman, S. A. & Johnsen, S. (1991). Coevolution to the edge of chaos: coupled fitness landscapes, poised states, and coevolutionary avalanches. *J Theor Biol.*, *21*, 149(4), 467-505. https://doi.org/10.1016/S0022-5193(05)80094-3.

Kauffman, S. & Levin, S. J. (1987). Towards a general theory of adaptive walks on rugged landscapes. *Theor Biol.*, *7*, 128(1), 11-45. https://doi.org/10.1016/S0022-5193%2887%2980029-2.

Kobayashi, H., Suzuki, T. & Unemoto, T. (1986). Streptococcal cytoplasmic pH is regulated by changes in amount and activity of a proton-translocating ATPase. *J Biol Chem.*, *261*, 627–630, 377.

Kobayashi, M., Kondo, S., Yasumoto, T. & Ohizumi, Y. (1986). Cardiotoxic effects of maitotoxin. A principal toxin of sea food poisoning on guinea pig and rat cardiac muscle. *J. Pharmacol. Exp. Ther.*, *238*, 1077-1083.

Koch, A. (1996). What size should a bacterium be? A question of scale. *Annu Rev Microbiol*, *50*, 317–348. https://doi.org/10.1146/annurev.micro.50.1.317.

Küchler, A., Yoshimoto, M., Luginbühl, S., Mavelli, F. & Walde, P. (2016). Enzymatic reactions in confined environments. *Nat Nanotechnol.*, *5*, 11(5), 409-20. https://doi.org/10.1038/nnano.2016.54.

Lammers, T. G. & Freeman, C. E. (1986). Ornithophily Among the Hawaiian Lobelioideae (Campanulaceae): Evidence from Floral Nectar Sugar Compositions. *American Journal of Botany*, *73*, 1613–1619. JSTOR, doi: 10.2307/2443929.

Li, X. Z. & Nikaido, H. (2009). Efflux-mediated drug resistance in bacteria: an update. *Drugs*, *69*, 1555–1623. https://dx.doi.org/10.2165%2F11317030-000000000-00000.

Lima Pérez, J., Rodríguez, D., Loera, O., Viniegra-González, G. & López-Pérez, M. (2017). Differences in Growth Physiology and Aggregation of *Pichia pastoris* Cells between Solid-State and Submerged Fermentations under Aerobic Conditions. *Journal of Chemical Technology & Biotechnology.*, *.93*, (2), 527-532 https://doi.org/10.1002/jctb.5384.

Maguire, M. E. (2006). Magnesium transporters: properties, regulation and structure. *Front. Biosci.*, *11*, 3149–63. http://dx.doi.org/10.2741/2039.

Maurer, L. M., Yohannes, E., Bondurant, S. S., Radmacher, M. & Slonczewski, J. L. (2005). pH regulates genes for flagellar motility, catabolism, and oxidative stress in *Escherichia coli* K-12. *J Bacteriol.*, 187, 304–319. doi: 10.1128/JB.187.1.304-319.

Miller, A. A., Engleberg, N. C. & DiRita, V. J. (2001). Repression of virulence genes by phosphorylation-dependent oligomerization of CsrR at target promoters in *S. pyogenes*. *Mol. Microbiol.*, *40*, 976–90. doi: 10.1046/j.1365-2958.2001.02441.x.

Moffatt, J., Harper, M., Harrison, P., Hale, J., Vinogradov, E., Seemann, T., Henry, R., Crane, Michael, F., Cox, A., Adler, B., Nation, R., Li, J. & Boyce, J. (2010). Colistin Resistance in *Acinetobacter baumannii* Is Mediated by Complete Loss of Lipopolysaccharide Production. *Antimicrobial agents and chemotherapy.*, *54*, 4971-7. https://dx.doi.org/10.1128%2FAAC.00834-10.

Munita, J. M. & Arias, C. A. (2016). Mechanisms of Antibiotic Resistance. *Microbiol Spectr.*, *4*(2), 0016-2015. doi: 10.1128/microbiolspec.VMBF-0016-2015.

Noguchi, K., Riggins, D. P., Eldahan, K. C., Kitko, R. D. & Slonczewski, J. L. (2010). Hydrogenase-3 contributes to anaerobic acid resistance of *Escherichia coli. PLoS One.*, *12*, 5(4), e10132. doi: 10.1371/journal.pone.0010132.

Ochs, M. M., McCusker, M. P., Bains, M. & Hancock, R. E. (1999). Negative regulation of the Pseudomonas aeruginosa outer membrane porin OprD selective for imipenem and basic amino acids. *Antimicrob Agents Chemother.*, *43*(5), 1085-90. PMCID: PMC89115.

Padan, E., Bibi, E., Ito, M. & Krulwich, T. (2005). Alkaline pH Homeostasis in Bacteria: New Insights. *Biochimica et biophysica acta.*, *1717*, 67-88. https://doi.org/10.1016/j.bbamem.2005.09.010.

Perry, J., Waglechner, N. & Wright. G. (2016). The Prehistory of Antibiotic Resistance. *Cold Spring Harbor perspectives in medicine*, *6*(6), a025197. doi: 10.1101/cshperspect.a025197.

Poehlsgaard, J. & Douthwaite, S. (2005). The bacterial ribosome as a target for antibiotics. *Nature reviews Microbiology.*, *3*, 870-81. 431 https://doi.org/10.1038/nrmicro1265.

Poole, K. (2000). Efflux-mediated resistance to fluoroquinolones in gram-negative bacteria. *Antimicrob Agents Chemother.*, *44*(9), 2233-41. https://dx.doi.org/10.1128%2Faac.44.10.2595-2599.

Qian, C. J., Yang, J., Yin, C. & Yulong, K. Y. (2019). Quorum Sensing: A Prospective Therapeutic Target for Bacterial Diseases *BioMed Research International*, (7), 1-15. https://doi.org/10.1155/2019/2015978.

Ramirez, M. & Tolmasky, M. (2010). Aminoglycoside Modifying Enzymes. *Drug resistance updates: reviews and commentaries in antimicrobial and anticancer chemotherapy*, *13*, 151-71. https://dx.doi.org/10.1016%2Fj.drup.2010.08.003.

Richter, O. M. H. & Ludwig, B. (2003). Cytochrome c oxidase structure, function, and physiology of a redox-driven molecular machine. *Rev. Physiol. Biochem. Pharmacol.*, *147*, 47–74. doi: 10.1007/s10254-003-0006-0.

Rousk, J. & Bååth, E. (2011). Growth of saprotrophic fungi and bacteria in soil, *FEMS Microbiology Ecology*, (78), *1*, 17–30. doi: 10.1111/j.1574-6941.2011.01106.x.

Silhavy, T. J., Kahne, D. & Walker, S. (2010). The bacterial cell envelope. *Cold Spring Harb Perspect Biol.*, *2*(5), a000414. doi: 10.1101/cshperspect.a000414.

Shrivastava, R. & Chang, S. S. (2019). Lipid trafficking across the Gram-negative cell envelope *The Journal of Biological Chemistry*, *294*, 14175-14184. https://dx.doi.org/10.1074%2Fjbc.AW119.008139.

Stancik, L., Stancik, D., Schmidt, B., Barnhart, D., Yoncheva, Y. & Slonczewski, J. (2002). pH-Dependent Expression of Periplasmic Proteins and Amino Acid Catabolism *in Escherichia coli*. *Journal of bacteriology.*, *184*, 4246-58. doi: 10.1128/JB.184.15.4246-4258.

Stern, D. (2013). The genetic causes of convergent evolution. *Nature reviews Genetics.*, *14*, 10.1038/nrg3483. doi: 10.1038/nrg3483.

Thom, G. & Prescott, C. D. (1997). The selection *in vivo* and characterization of an RNA recognition motif for spectinomycin *Bioorg. Med. Chem*, *5*, 1081-1086. https://doi.org/10.1016/S0968-0896(97)00060-6.

Thompson, J. N., Nuismer, S. L. & Gomulkiewicz, R. (2002). Coevolution and Maladaptation, *Integrative and Comparative Biology*, *42*, (2), 381–387. doi: 10.1093/icb/42.2.381.

Torday, J. S. (2015). Homeostasis as the Mechanism of Evolution. *Biology*, *4*(3), 573–590. https://dx.doi.org/10.3390%2Fbiology4030573.

Tsurusako, K. F., Nishizaki, K., Takata, T., Ogawa, T., Nakashima, T., Sugata, K. N., Yorizane, S., Ogawara, T. & Yuichi, Y. M. (1999). The

Alteration of Penicillin-Binding Proteins (PBPs) in Drug-resistant *Streptococcus pneumoniae* Isolated from Acute Otitis Media *Acta Oto-Laryngologica*, *119*, (540). https://doi.org/10.1080/00016489950181233.

Tyukina, T., Smirnova, E. & Pokidysheva, L. (2016). Evolution of adaptation mechanisms: Adaptation energy, stress, and oscillating death. *Journal of Theoretical Biology*, *405*, 127-139. https://doi.org/10.1016/j.jtbi.2015.12.017.

Viniegra-González, G. & Favela-Torres, E. (2005). Why Solid-State Fermentation Seem to be Resistant to Catabolite Repression? *Food Technol. Biotechnol.*, *44*, 397-406 https://hrcak.srce.hr/109941.

Wisotzkey, J. D., Jurtshuk, P., Jr. Fox, G. E., Deinhard, G. & Poralla, K. (1992). Comparative sequence analyses on the 16S rRNA (rDNA) of *Bacillus acidocaldarius*, *Bacillus acidoterrestris*, and *Bacillus cycloheptanicus* and proposal for creation of a new genus, *Alicyclobacillusgen. Int J Syst Bacteriol.*, *42*, 263–269. doi: 10.1099/00207713-42-2-263.

Wright, G. D. (2005). Bacterial resistance to antibiotics: enzymatic degradation and modification. *Adv Drug Deliv Rev.*, *29*, 57(10), 1451-70. https://doi.org/10.1016/j.addr.2005.04.002.

Wright, S. (1939). Statistical Genetics in Relation to Evolution. *Exposés de Biométrie et de la statistique biologique XIII*. Hermann & Cie, Paris. Reprinted in W. B. Provine Sewall Wright: Evolution: Selected Papers. University of Chicago Press, Chicago, IL, 1986, 283–341.

Yin, W., Wang, Y., Liu, L. & He, J. (2019). Biofilms: The Microbial "Protective Clothing" in Extreme Environments. *International journal of molecular sciences*, *20*(14), 3423. https://dx.doi.org/10.3390%2Fijms20143423.

Zhang, X., Chen, Q. & Han, X. (2013). Soil bacterial communities respond to mowing and nutrient addition in a steppe ecosystem. *PloS one*, *8*(12), e84210. https://doi.org/10.1371/journal.pone.0084210.

Zotter, A., Bäuerle, F., Dey, D., Kiss, V. & Schreiber, G. (2017). Quantifying enzyme activity in living cells *J. Biol. Chem.*, *292*, 15838. doi: 10.1074/jbc.M117.792119.

In: Molecular Basis of Specific Mechanism … ISBN: 978-1-53618-751-9
Editors: Marcos López-Pérez et al. © 2020 Nova Science Publishers, Inc.

Chapter 2

MOLECULAR MECHANISM IMPLICATED IN CONIDIA PRODUCTION BY ENTOMOPATHOGEN FUNGI

Divanery Rodríguez-Gómez[*]
Coordinación de Ingeniería Bioquímica,
Instituto Tecnológico Superior de Irapuato, Guanajuato, México

ABSTRACT

Microbial ecology in agricultural systems are of special relevance because they allow the development of technologies to improve yields and different qualities of this economic sector. In the last decades, there are microorganisms that have been used in different biological control techniques, using a natural antagonism approach that makes it possible to control or mitigate the impact of pest-causing etiological agents. In this chapter, the molecular bases that explain the production processes of spores of entomopathogenic fungi are explored.

Keywords: conidia, entomopathogenic fungi

---

[*] Corresponding Author's Email: divanery.rodriguez@itesi.edu.mx.

1. INTRODUCTION

Beauveria bassiana and *Metarhizium* spp. are entomopathogen fungi found worldwide in the soil, they have been produced on a large scale and have a spectrum of control that includes egg, adults and larvae of many kinds of insects; such as moths, crickets, ants, beetle, weevils, whiteflies, grasshoppers, thrips, aphids, mites and many others. Entomopathogen fungi that belong to the subdivision Hypocreales are filamentous fungi with a simple life cycle with conidia production during its asexual stage. It is the infective unit that attaches to the insect cuticle and should have some important features then the pathogenicity would be appropiate in the field. Therefore, extensive research have been related to the mechanisms used by fungi to increase conidiation in culture, even at industrial level.

Because of their primary role in the infection process, conidia have been considered from the beginning of biological control history as the most adapted fungal propagule to produce. Conidia are produced in solid state culture meanwhile blastospores are produced in submerged conditions, they have different features mainly related to surface characteristics, but also to enzyme content and release (Lopez et al., 2015).

Regarding to strategies for conidial production increase in entomopathogenic fungi it is known that most of the physiological factors involved are nutritional and environmental, like substrate composition, C:N relationship, type of culture (Lopez et al., 2015; Rodriguez-Gomez et al., 2009), concentration and flux of oxygen (Tlecuitl-Beristain et al., 2010; Garza-Lopez et al., 2012; Miranda-Hernandez et al., 2013; Rodriguez-Gomez et al., 2016; Garcia-Ortiz, 2015), light regimes and types (Zhang et al., 2009; Sanchez Murillo et al., 2004) among others. In other fungi that are not entomopathogenic, such as *Trichoderma* spp. physical stimuli, such as light exposure and C:N relation are known to trigger conidiation, together with others like the ambient pH, extracellular calcium and mechanical injury to the mycelium; however, conidiogenesis itself is a holistic response determined by the cells metabolic state.

Conidiation is a survival mechanism of the mycelia when convenient conditions are not possible, little is known about genes implicated in regulation of conidia production, besides, not everything can be elucidated since some are cascades of regulation, where one gene is altered and results in multi-phenotypic defects, having as intermediates pivotal regulation, such as cAMP and PKA. In cases of nutritional, osmotic, or oxidative stress, it can be related to generation of reactive oxygen species (ROS), and therefore to conidia production. Besides, the findings on regulation of genes toward conidia production has showed important cross talk which is the acquired resistance to some stressful conditions related to a different stress condition of growth, This cross talk may be related to mechanism that are related to oxidative stress production, since some of the cascades action of the genes can develop the on/off in genes related to regulate the response toward oxidative stress (enzymes mainly).

Conidiation and stress tolerance are two important factors that could determine the success of entomopathogenic fungi for insect pest management, they are related and are dependant on the production conditions. Therefore understanding the molecular mechanisms of conidia production have implications for commercial development in order to obtain more tolerant to field stressful conditions such as temperature, UV light, water availabilty, to obtain a full biocontrol potential.

2. SIGNAL REGULATION NETWORKS

A complex network of signal transduction pathways including cAMP-PKA, MAPK, G protein signalling cascades play a pivotal role in conidiation of filamentous fungi, such network can be triggered by several stress conditions, such as nutrient deprivation. cAMP activates protein kinase A (PKA), resulting in phosphorylation of proteins in a regulatory cascade that would give place to signal at nucleic level which in return would activate transcription factors related to conidiation. MAPK genes are in some cases required but in other cases are totally dispensable for conidiation (Carreras-Villaseñor et al., 2012), depending on the stimulus and its specific regulatory network. Besides, in *Aspergillus nidulans* some regulatory genes like *brlA, abaA, wetA* and *flbB, flbC, flbD, flbE* define the central regulatory pathway that controls conidiation-specific gene expression, some orthologue genes have been found in other fungi like *Neurospora crassa* (Adams et al., 1998), *Metarhizium robertsii*, *Metarhizium anisopliae*, *Beauveria bassiana*, *Trichoderma virens*, among others.

3. G PROTEIN REGULATION

In filamentous fungi, heterotrimeric G protein signalling pathways are involved in conidiation, mating, pathogenicity, secondary metabolite production (Brodhagen and Keller, 2006), fungal development, stress response and vegetative incompatibility (Fang et al., 2007; Yu et al., 2006).

The binding of a ligand to a G protein-coupled receptor (GPCR) is the event that activates G protein signalling. GPCRs reside in the cell membrane, where they perceive extracellular signals such as light, ions, amino acids, sugars, nucleic acids, steroids, polypeptides and fatty acids, and transduce this information about the external environment across the membrane to heterotrimeric G proteins comprising Gα, Gβ and Gγ subunits. Upon ligand binding, G-protein-coupled receptors (GPCRs) catalyse the exchange of GDP to GTP on the α-subunit. While dissociated, Gα and/or the G$\beta\gamma$ complex relay messages to other downstream effectors in fungi; these second messenger pathways are primarily (i) MAP kinase protein phosphorylation cascades and (ii) adenylyl cyclase/cAMP/PKA pathways. The activation is terminated by GTP hydrolysis by the GTPase domain of the α-subunit, leaving the α-subunit in the GDP-bound form and allowing heterotrimerization to resume (Hamm, 1998).

The signal amplitude of G protein signalling is determined by the balance of the rates of GDP/GTP exchange (activation) and the rates of GTP hydrolysis (deactivation), which is achieved by the intrinsic GTPase of the Gα subunit. However, the GTPase of the Gα subunit requires a regulator of G protein signalling (RGS) to enhance GTPase activity (Ross & Wilkie, 2000). RGS protein families (orthologues of *RgsA*, *RgsB*, *RgsC*, *RgsD*, *FlbA* and *Gprk*) have been identified in *Aspergillus* spp. and other fungi. *FlbA* negatively regulates signalling for vegetative growth and activated conidiation, mediated through the Ga subunit, FadA (Yu et al., 1996), and the Gb subunit, SfaD (Rosen et al., 1999). Han et al. (2004) described a second RGS protein in *A. nidulans*, RgsA, which down-regulated stress responses and stimulated conidiation through attenuation of GanB (Ga) signalling.

RGS protein gene, *cag8*, from *M. anisopliae* plays an important role in the regulation of conidiation, virulence and hydrophobin synthesis as described by Fang et al. (2008), it showed significant homology to *FlbA* found in *Aspergilli*. In *B. bassiana* the gene *bbrgs1* transcription responds to a posible oxidative stress mediated by higher concentration of O_2 in the air flux (Rodriguez-Gomez et al., 2016). Besides, the homologous gene found in *Beauveria bassiana* and *Metarhizium anisopliae* (GPR-1) that activates heterotrimeric G proteins senses different carbon sources, and is involved in conidia production (Fang et al., 2008).

3.1. Ras1 and Ras2 Proteins

Ras1 and Ras2 are two distinct Ras GTPases in *Beauveria bassiana*, this kind of proteins stimulates kinases like MAP kinases in order to regulate gene transcription, they cycle between GTP-bound (active) and GDP-bound (inactive) conformations and act as switches in the signalling hub of molecular circuits. They regulate differentially the germination, growth, conidiation, multi-stress tolerance and virulence of *B. bassiana*. Xie et al. (2013) found that both Ras1 and Ras2 in *B. bassiana* have proved to be essential for regulating cell tolerances to oxidation, cell wall disturbance and hyperosmolarity during colony growth and conidial germination which also affects biocontrol potential of *B. bassiana*. Such importance is evident in *Ras1* and *Ras2* mutants being more sensitive to oxidation or cell wall disturbance might become less capable of detoxifying superoxide anions from infected host cells or protecting cells from external damage (Xie et al., 2013).

4. LIPIDIC MOLECULES

Oxylipins (oxygenated fatty acid-derived molecules implicated as intra- and intercellular signals) are tied to G protein signalling pathways in fungi. G protein signalling is initiated when an appropriate GPCR at the cell surface undergoes ligand perception. Mutations in genes for oxylipin biosynthesis result in aberrant transcription of genes governing sporulation and secondary metabolism like *brlA* (Brodhagen and Keller, 2006). In *Aspergillus* and *Fusarium* oxylipins signals and pathogenic mechanism have been asocciated through cutin degradation by lipase enzymes called cutinases. On the other hand, for plant pathogens VOCs which are the end products of fatty acid metabolism and peptaibols stimulate conidiation (Carreras-Villaseñor et al., 2012). Meanwhile, entomopathogenic fungi also include enzyme production like chitinases, proteases and lipases during its pathogenic phase. The production of pr1 subtilisin like protein in *M. anisopliae* during its conidiation process have been related to its conidiation process (Cherrie Lee and Bidochka, 2005). Some specific lipoperoxidation or oxidized proteins generated during the oxidative stress by an oxygen flux have been found related to changes in conidiation in *B. bassiana* (Garza-Lopez et al., 2012). However, further research on the implications of the metabolites from proteic or lipidic nature must be addressed.

5. CALCIUM RELATED PROTEINS

In fungi, calcium plays important roles during differentiation to induce conidiation independently of the nutritional state. P-type Ca^{2+}-ATPase (Pmr1) is a core element in calcium-calcineurin pathway and evidence for its cross-talk with other signalling pathways in filamentous fungi. Wang et al. (2013) reported its importance in *Beauveria bassiana*. Its inactivation caused severe defects sensitivity to almost all tested types of stressful chemicals during colony growth and conidial germination, drastic reductions in cell tolerances to oxidative, hyperosmotic, cell wall disturbing and fungicidal stresses and toxic metal ions during colony growth and/or a loss of ~71% conidial yield after 7-day cultivation under normal conditions, and half loss of the fungal biocontrol potential represented by conidial virulence, thermotolerance and UV-B resistance. Therefore Bbpmr is not only a regulator of Ca^{2+}/Mn^{2+} homeostasis but also of various processes that has multi-stress responses through cross-talk with cellular signalling networks in response to nutritional, chemical and environmental stresses.

6. LIGHT REGULATION

B. bassiana is effectively stimulated by white light and blue light, blue light was the most effective to stimulating conidiation. *B. bassiana* requires a certain growth period to gain photoadaptation in terms of conidiation needed a relatively longer growth time (96 h) when compared to *Trichoderma, Phycomyces* and *Paecilomyces*, they required 16, 48, and 72 h growth time for the response, respectively (Zhang et al., 2009). In model fungal systems it has been found that phototropism, resetting of the circadian rhythm, the induction of carotenogenesis and the development of reproductive structures are controlled by blue light. It was also reported for *Isaria*, another entomopathogenic fungus, that its growth in continuous illumination or under a night-day regime resulted in prolific formation of conidiophores bearing abundant mature conidia (Sánchez-Murillo, et al., 2004).

The effect of light on conidiation have been widely studied in *Trichoderma atroviridae* and *T. reesei*, a conserved mechanism of environmental perception through the White Collar orthologues blue light receptors (BLR-1 and BLR-2) have been documented. More than 40 genes have been shown to be regulated by BLR-1/BLR-2 (Esquivel-Naranjo & Herrera-Estrella, 2007). VIVID, a small PAS/LOV domain protein, acts as the blue-light photoreceptor. Orthologue of VIVID, designated ENVOY, has been found in *Hypocrea jecorina* (*T. reesei*) showing to regulate photoadaptation of BLR-1/BLR-2 and has also been found in *Beauveria bassiana* (BBA_02876, NCBI). The BLR-1 and BLR-2 proteins appears to play an essential role as a sensor and transcriptional regulator in photoconidiation. BLR proteins are necessary for carbon deprivation induced conidiation, even in the absence of light, pointing to the existence of an unprecedented cross talk between light and carbon sensing. PKA plays an important role in the regulation of light responses in *Trichoderma*. It is noteworthy that light induces oxidative stress evidenced by Transcriptomic analyses (Casas-Flores et al., 2006).

7. VELVET MECHANISM

The VELVET protein is a global regulator of morphogenesis and secondary metabolism in filamentous fungi. Deletion of *vel1* in *Trichoderma virens* results in a total loss of conidiation, in *Aspergillus nidulans* and *Neurospora crassa* orthologue genes have been found related to cell differentiation induced primarily by desiccation, carbon source limitation as well as light and changes in CO_2 partial pressure (Bayram et al., 2008) Orthologue genes for *Metarhizium robertsii, Metarhizium acridum* and *Beauveria bassiana* are found in gene databases (NCBI). The deletion of *VeA* gene in *B. bassiana* resulted in facilitated hyphal growth and decreased cell length on rich media, light growth defects and increased sensitivities to oxidation, high osmolarity and prolonged heat shock during colony growth. Compared to wild-type, the deletion affected multistress response mediated by light but no virulence parameters (Wang et al., 2019).

7.1. Injury and ROS Production

In *B. bassiana*, atmospheric modification (16 and 26% O_2) resulted in more rapid conidia formation at early stages following O_2 modification. The maximum yield of conidia was achieved under hypoxia (16% O_2) and was related to increase in the activity of both enzymes superoxide dismutase and catalase (Garza-Lopez et al., 2011). Reactive oxygen species (ROS) are byproducts of normal aerobic metabolism, produced mainly by partial reduction of oxygen during respiration. Such molecules can oxidize virtually any cell molecule, causing DNA damage, protein inactivation, protein cross-linking and fragmentation, and lipid peroxidation. Antioxidant responses involve a prokaryotic-type multistep phosphorelay coupled to a stress-response MAP kinase pathway and an AP-1 type transcription factor. Filamentous fungi have the presence of a larger number of phosphorelay sensor kinases, antioxidant enzymes (dismutate, catalases, peroxidases, glutathione peroxidases and peroxiredoxins) and secondary metabolites with antioxidant functions.

NADPH enzyme complex to produce ROS, which combined with nitric oxide (NO^+) produce the nitrogen reactive species peroxynitrite (Nox). During early stages of the response to injury, genes known to generate ROS are induced, whereas those known to scavenge ROS are repressed. Light and mechanical injury activate the Nox complex for the production of ROS, leading to the oxidation of proteins and lipids. In Fungi *NoxA* (*Nox1*) and *NoxB* (*Nox2*), and *NoxC, NoxR*, play a key role in fungal cell differentiation and development (Carreras-Villaseñor et al., 2012), *NoxR* can be found in *B. bassiana, M. acridum* and *M. robertsii* (NCBI).

8. OTHER IMPORTANT GENES

Some other genes related to different mechanisms in the cell regulation have been studied to be important at some extent in the fungal conidiation. The gene encoding b-1,3-glucan synthase was reported for the entomopathogenic fungus *Metarhizium acridum* (MaFKS; HQ441252) as an essential cell wall structural component important for maintenance of cell wall elasticity in fungi hyperosmotic pressure tolerance and conidiation (Yang et al. 2011).

A Group III histidine kinase (mhk1) upstream of high-osmolarity glycerol pathway regulates sporulation, in *Metarhizium robertsii* the *mhk1* disruption increased tolerance to H_2O_2, to hyperosmolarity, thermotolerance and increased conidial yield in 50-67%, also decreased conidial UV-B resistance and virulence toward *T. molitor* larvae. The *mhk1* disruption elevated the transcripts of nine genes, including two associated with conidiation (*flbC* and *hymA*). Then again a great complexity in mediating the fungal multi-stress response and the conidiation is evidenced (Zhou et al., 2012).

B. bassiana, JEN1 (*BbJEN1*) homologous gene to *JEN1* encoding a carboxylate transporter which is a lactate/pyruvate symporter in *Saccharomyces cerevisiae* was disrupted resulting in decreased carboxylate contents in hyphae, while increasing the conidial yield. However, overexpression of this transporter resulted in significant increases in carboxylates and decreased the conidial yield. Besides, *BbJEN1* was highly expressed in the hyphae penetrating insect cuticles suggesting that this gene is also involved in the early stage of pathogenesis of *B. bassiana* (Jin et al., 2010).

CONCLUSION

Environmental and nutritional culture conditions affect the growth, development of conidia, and performance of the conidia in different scenarios such as store, field and on the insect. In order to produce stress tolerant conidia which are less affected by low humidity, UV-light, elevated temperatures and fungicides and to have more virulence, better understanding of mechanism of conidiation could help improve homogeneity in conidia production process to obtain quality parameters controlled.

Here, some specific genes and regulation pathways were reviewed as being relevant in conidia production for species of entomopathogenic fungi *Beauveria* and *Metarhizium*. We found in all cases some stress induction as a trigger to conidiation and the cross-talk mechanism implicated as an advantage toward the great catalogue of factors affecting conidial performance.

REFERENCES

Adams, T. H., Wieser, J. K., & Yu, J. H. (1998). Asexual sporulation in *Aspergillus nidulans*. *Microbiol. Mol. Biol. Rev.*, 62(1), 35-54. PMCID: PMC98905.

Bayram, Ö., Krappmann, S., Seiler, S., Vogt, N., & Braus, G. H. (2008). *Neurospora crassa* ve-1 affects asexual conidiation. *Fungal Genetics and Biology*, 45(2), 127-138. https://doi.org/10.1016/j.fgb.2007.06.001.

Brodhagen, M., & Keller, N. P. (2006). Signalling pathways connecting mycotoxin production and sporulation. *Molecular Plant Pathology*, 7(4), 285-301. doi: 10.1111/j.1364-3703.2006.00338.x.

Carreras-Villaseñor, N., Sánchez-Arreguín, J. A., & Herrera-Estrella, A. H. (2012). *Trichoderma*: sensing the environment for survival and dispersal. *Microbiology*, 158(1), 3-16. doi: 10.1099/mic.0.052688-0.

Casas-Flores, S., Rios-Momberg, M., Rosales-Saavedra, T., Martínez-Hernández, P., Olmedo-Monfil, V., & Herrera-Estrella, A. (2006). Cross talk between a fungal blue-light perception system and the cyclic AMP signaling pathway. *Eukaryotic Cell*, 5(3), 499-506. doi: 10.1128/EC.5.3.499-506.2006.

Esquivel-Naranjo, E. U., & Herrera-Estrella, A. (2007). Enhanced responsiveness and sensitivity to blue light by blr-2 overexpression in *Trichoderma atroviride*. *Microbiology*, 153(11), 3909-3922. doi: 10.1099/mic.0.2007/007302-0.

Fang, W., Pei, Y., & Bidochka, M. J. (2007). A regulator of a G protein signalling (RGS) gene, cag8, from the insect-pathogenic fungus *Metarhizium anisopliae* is involved in conidiation, virulence and hydrophobin synthesis. *Microbiology*, 153(4), 1017-1025. doi: 10.1099/mic.0.2006/002105-0.

Fang, W., Scully, L. R., Zhang, L., Pei, Y., & Bidochka, M. J. (2008). Implication of a regulator of G protein signalling (BbRGS1) in conidiation and conidial thermotolerance of the insect pathogenic fungus Beauveria bassiana. *FEMS Microbiology Letters*, 279(2), 146-156. doi: 10.1111/j.1574-6968.2007.00978.x.

Garcia-Ortiz, N., Tlecuitl-Beristain, S., Favela-Torres, E., & Loera, O. (2015). Production and quality of conidia by *Metarhizium anisopliae var. lepidiotum*: critical oxygen level and period of mycelium competence. *Applied Microbiology and Biotechnology*, 99(6), 2783-2791. doi: 10.1007/s00253-014-6225-2.

Garza-López, P. M., Konigsberg, M., Gómez-Quiroz, L. E., & Loera, O. (2012). Physiological and antioxidant response by *Beauveria bassiana* Bals (Vuill.) to different oxygen concentrations. *World Journal of Microbiology and Biotechnology*, 28(1), 353-359. doi: 10.1007/s11274-011-0827-y.

Hamm, H. E. (1998). The many faces of G protein signaling. *Journal of Biological Chemistry*, 273(2), 669-672. doi: 10.1074/jbc.273.2.669.

Han, K. H., Seo, J. A., & Yu, J. H. (2004). Regulators of G-protein signalling in *Aspergillus nidulans*: RgsA downregulates stress response and stimulates asexual sporulation through attenuation of GanB (Gα) signalling. *Molecular Microbiology*, 53(2), 529-540. https://doi.org/10.1111/j.1365-2958.2004.04163.x.

Jin, K., Zhang, Y., Fang, W., Luo, Z., Zhou, Y., & Pei, Y. (2010). Carboxylate transporter gene JEN1 from the entomopathogenic fungus *Beauveria bassiana* is involved in conidiation and virulence. *Appl. Environ. Microbiol.*, 76(1), 254-263. doi: 10.1128/AEM.00882-09.

Lopez-Perez, M., Rodriguez-Gomez, D., & Loera, O. (2015). Production of conidia of *Beauveria bassiana* in solid-state culture: current status and future perspectives. *Critical Reviews in Biotechnology*, 35(3), 334-341. doi: 10.3109/07388551.2013.857293.

Miranda-Hernández, F., Saucedo-Castañeda, G., Alatorre-Rosas, R., & Loera, O. (2014). Oxygen-rich culture conditions enhance the conidial infectivity and the quality of two strains of *Isaria fumosorosea* for potentially improved biocontrol processes. *Pest Management Science*, 70(4), 661-666. doi: 10.1002/ps.3605.

Rodríguez-Gómez, D., Loera, O., Saucedo-Castañeda, G., & Viniegra-González, G. (2009). Substrate influence on physiology and virulence of *Beauveria bassiana* acting on larvae and adults of Tenebrio molitor.

World Journal of Microbiology and Biotechnology, 25(3), 513-518. https://doi.org/10.1007/s11274-008-9917-x.

Rodriguez-Gomez, D., Marcial-Quino, J., & Loera, O. (2015). Modulation of conidia production and expression of the gene bbrgs1 from *Beauveria bassiana* by oxygen pulses and light. *Journal of Invertebrate Pathology*, 130, 82-87. https://doi.org/10.1016/j.jip.2015.07.004.

Rosén, S., Yu, J. H., & Adams, T. H. (1999). The *Aspergillus nidulans* sfaD gene encodes a G protein β subunit that is required for normal growth and repression of sporulation. *The EMBO Journal*, 18(20), 5592-5600. https://doi.org/10.1093/emboj/18.20.5592.

Ross, E. M., & Wilkie, T. M. (2000). GTPase-activating proteins for heterotrimeric G proteins: regulators of G protein signaling (RGS) and RGS-like proteins. *Annual Review of Biochemistry*, 69(1), 795-827. https://doi.org/10.1146/annurev.biochem.69.1.795.

Sanchez-Murillo, R. I., de la Torre-Martínez, M., Aguirre-Linares, J., & Herrera-Estrella, A. (2004). Light-regulated asexual reproduction in *Paecilomyces fumosoroseus*. *Microbiology*, 150(2), 311-319. doi: 10.1099/mic.0.26717-0.

Cherrie-Lee, N., & Bidochka, M. J. (2005). Up-regulation of Pr1, a subtilisin-like protease, during conidiation in the insect pathogen *Metarhizium anisopliae*. *Mycological Research*, 109(3), 307-313. https://doi.org/10.1017/S0953756204001856.

Tlecuitl-Beristain, S., Viniegra-González, G., Díaz-Godínez, G., & Loera, O. (2010). Medium selection and effect of higher oxygen concentration pulses on *Metarhizium anisopliae var. lepidiotum* conidial production and quality. *Mycopathologia*, 169(5), 387-394. doi: 10.1007/s11046-009-9268-7.

Wang, D. Y., Tong, S. M., Guan, Y., Ying, S. H., & Feng, M. G. (2019). The velvet protein VeA functions in asexual cycle, stress tolerance and transcriptional regulation of *Beauveria bassiana*. *Fungal Genetics and Biology*, 127, 1-11. https://doi.org/10.1016/j.fgb.2019.02.009.

Wang, J., Zhou, G., Ying, S. H., & Feng, M. G. (2013). P-type calcium ATPase functions as a core regulator of *Beauveria bassiana* growth, conidiation and responses to multiple stressful stimuli through cross-talk with signalling networks. *Environmental Microbiology*, 15(3), 967-979. doi: 10.1111/1462-2920.12044.

Xie, X. Q., Guan, Y., Ying, S. H., & Feng, M. G. (2013). Differentiated functions of Ras1 and Ras2 proteins in regulating the germination, growth, conidiation, multi-stress tolerance and virulence of *Beauveria bassiana*. *Environmental Microbiology*, 15(2), 447-462. doi: 10.1111/j.1462-2920.2012.02871.x.

Yang, M., Jin, K., & Xia, Y. (2011). MaFKS, a β-1, 3-glucan synthase, is involved in cell wall integrity, hyperosmotic pressure tolerance and conidiation in *Metarhizium acridum*. *Current Genetics*, 57(4), 253-260. doi: 10.1007/s00294-011-0344-4.

Yu, J. H., Wieser, J., & Adams, T. H. (1996). The Aspergillus FlbA RGS domain protein antagonizes G protein signaling to block proliferation and allow development. *The EMBO Journal*, 15(19), 5184-5190. PMC452262.

Yu, J. H., Mah, J. H., & Seo, J. A. (2006). Growth and developmental control in the model and pathogenic aspergilli. *Eukaryotic Cell*, 5(10), 1577-1584. https://doi.org/10.1128/EC.00193-06.

Zhang, Y. J., Li, Z. H., Luo, Z. B., Zhang, J. Q., Fan, Y. H., Pei, Y. 2009. Light stimulates conidiation of the entomopathogenic fungus *Beauveria bassiana*. *Bio. Sci. and Tech.* 19:91-101 https://doi.org/10.1080/09583150802588516.

Zhou, G., Wang, J., Qiu, L., & Feng, M. G. (2012). A Group III histidine kinase (mhk1) upstream of high-osmolarity glycerol pathway regulates sporulation, multi-stress tolerance and virulence of *Metarhizium robertsii*, a fungal entomopathogen. *Environmental Microbiology*, 14(3), 817-829. doi: 10.1111/j.1462-2920.2011.02643.x.

In: Molecular Basis of Specific Mechanism ...   ISBN: 978-1-53618-751-9
Editors: Marcos López-Pérez et al.   © 2020 Nova Science Publishers, Inc.

Chapter 3

MOLECULAR BASIS OF SPECIFIC STRATEGIES USED BY MICROORGANISMS TO COPE WITH STRESS: THE CASE OF *STREPTOMYCES*

Hypatia Arano-Varela[*] *and Francisco J. Fernández*[†]
Department of Biotechnology,
Universidad Autónoma Metropolitana-Iztapalapa.
Mexico City, Mexico

ABSTRACT

Life on earth is a challenge. Living organisms have been forced to adapt to constant environmental changes. They also had to learn to live in society within and among species and to defend themselves to survive and ensure the prevalence of their genes for the next generations. A smart and successful tool to achieve this has been the development of biochemical mechanisms, deriving in the biosynthesis of natural products (known as secondary or 'specialized' metabolites). Through these chemical compounds, organisms react to changes in their

---

[*] Corresponding Author's Email: ypatya@gmail.com.
[†] Corresponding Author's Email: fjfp@xanum.uam.mx.

physicochemical and biological environment. In this chapter, we aimed to illustrate the intricate molecular mechanisms implicated in transducing the environmental signals into metabolic outputs, as well as to describe novel discoveries by choosing as case in point the stress induced by oxidants and the translation-blocking antibiotics in *Streptomyces*.

Keywords: environmental conditions, microbial natural products, production of antibiotics, sigma factors, signal transduction, stress, transcriptional factors

INTRODUCTION

Life on earth is a challenge. Living organisms have been forced to adapt to constant environmental changes (temperature, pH, oxygen content, oxidative stress, limited access to carbon, nitrogen, phosphate, exposure to toxins and antibiotic agents). They also had to learn to live in society within and among species and to defend themselves to survive and ensure the prevalence of their genes for the next generations.

A smart and successful tool to achieve this has been the development of biochemical mechanisms, deriving in the biosynthesis of natural products (known as secondary or 'specialized' metabolites). Their production has been reported as derived from diverse sources including plants, animals, microorganisms, marine organisms, among others (Pham et al., 2019).

Through these chemical compounds, organisms react to changes in their physicochemical and biological environment. Additionally, they are largely responsible for the microbial communication and interaction with other living organisms (Bérdy, 2012; Cornforth and Foster, 2013). Besides their ecological functions, these structurally and chemically diverse compounds represent a relevant source of modern human pharmaceutical compounds due to their various bioactivities, including antibiotic, antifungal, antiviral, anthelminthic, anticancer, immunosuppressant, anti-inflammatory, anti-hypertensive, anti-hypercholesterolemic, and biofilm inhibitory effects (Omura et al., 2001; Traxler and Kolter, 2015; Pham et al., 2019). These compounds are also extensively used in veterinary medicine, agronomy (herbicides and pesticides), and the food industry (food additives) (Bérdy, 2012).

Most of the known bacterial bioactive compounds come from soil microbial inhabitants. Among these, the phylum Actinobacteria is remarkable, as a diverse group of G+C (more than 70%), and Gram-positive bacteria known as 'actinomycetes' (a term conventionally used to describe any filamentous, Gram-positive actinobacterium from the soil). They are saprophytic, soil-dwelling bacteria that spend their life cycles mostly as spores, especially under nutrient scarcity circumstances, and represent an overwhelming source of biologically active compounds, especially bacteria of the genus *Streptomyces* (Nett et al., 2009; Traxler and Kolter, 2015; Barka et al., 2016).

Due to their significant industrial importance, several studies have been conducted to understand the molecular bases behind their morphological differentiation and the biosynthesis of natural products. To date, genetic engineering approaches (i.e., generation of mutants carrying deletions and/or overexpressing genes) along with next-generation sequencing (i.e., whole-genome sequencing, RNA-seq, ChIP-seq) and computational tools have greatly increased the knowledge regarding the main molecular processes (replication, transcription, translation) in cells. We have access now to a constantly growing and overwhelming amount of information regarding the activities and physiology of multiple molecules (including sigma factors, anti-sigma factors, anti-anti-sigma factors, second messengers, transcriptional factors, global regulators, among others); in addition, genomic information regarding transcriptional units architecture (regulatory elements, promoters, ribosome binding sites, 5'-UTRs, transcriptional terminators), transcriptomes, translatomes, proteomes, and metabolomes (to mention a few) is now at our reach. These data have allowed us to delineate the complex network of metabolic pathways involved in morphological and physiological microbial development (Sun et al., 2017; Park et al., 2019; Šmídová et al., 2019; Lee et al., 2020; Makitrynskyy et al., 2020), opening the possibility to improve the production of antibiotics. These antibiotics can be obtained now under normal laboratory conditions, providing a useful alternative to combat the increasing rise of antibiotic-resistant pathogens (Hwang et al., 2019; Belknap et al., 2020; Lee et al., 2020).

Usually, genes that encode the information for the biosynthesis of a specific natural product (i.e., antibiotics) are grouped in clusters, referred as 'secondary metabolism biosynthetic gene clusters' (smBGCs). The genome of model streptomycete *Streptomyces coelicolor* possesses more than 20 smBGCs, genomes of other model Actinobacteria species revealed a similar case with some species harboring more than 50 different smBGCs. However, most of them remain silent under laboratory growth conditions (Bentley et al., 2002; van der Heul et al., 2018).

Instead of encoding enzymes needed for catalyzing antibiotics production, smBGC clusters also harbor resistance genes, as well as genes with regulatory functions, generally termed 'cluster-situated regulators' (CSRs) which control the level of transcription of its associated antibiotic, which closely correlates with levels of antibiotic biosynthetic production (van der Heul et al., 2018; Wei et al., 2018). Additionally, there also exist many other regulators that control the biosynthetic process at higher levels (i.e., transcriptional factors). Pleiotropic regulators are located out of the BGCs, they control both morphological development and secondary metabolite biosynthesis. Global regulators are distributed all over the chromosome and control CSR genes, pleiotropic regulatory genes and central metabolic genes (Wei et al., 2018).

Typically, each smBGCs contains one or more CSR, among the most studied are ActII-ORF4 (controls actinorhodin, Act), RedD (controls undecylprodigiosin, Red), CdaR (controls calcium-dependent antibiotic, Cda) of the model actinomycete *S. coelicolor* and StrR (controls streptomycin). The presence of many CSRs is a common event, with each CSR controlling a subset of genes or taking part in a layered cascade (van der Heul et al., 2018; Wei et al., 2018). In the opposite scenario, the absence of CSR has been reported (Makitrynskyy et al., 2020; Yan et al., 2020).

Interactions between CSRs and pleiotropic or global regulators constitute the basis of regulation of antibiotic biosynthesis and also morphological differentiation; however, there exist many other regulatory factors at different levels that provide important checkpoints at distinct stages of control. Among others, sigma factors, anti-sigma factors and anti-anti-sigma factors (Sun et al., 2017; Park et al., 2019; Smidova et al., 2019); small nucleotide signals (van der Heul et al., 2018) and nucleotide second messengers (Latoscha et al., 2019; Makitrynskyy et al., 2020) all of which exert their function at transcriptional level. *S. coelicolor* genome harbors 700 regulatory genes (Bentley et al., 2002) mostly involved in sensing environmental variations (i.e., temperature, pH, oxygen content, oxidative stress, limited access to carbon, nitrogen, phosphate, exposure to toxins and antibiotic agents) and generating responses in order to survive by readjusting morphological and physiological processes.

The set of regulatory molecules results in a highly complex regulatory network coordinated at different hierarchical levels that control morphological development, growth rates, and natural product biosynthesis. Understanding the regulation of antibiotic biosynthesis opens the possibility to improve production of antibiotics that cannot be obtained under normal laboratory conditions, providing a useful alternative to combat the increasingly rise of antibiotic-resistant pathogens (Wei et al., 2018; Hwang et al., 2019; Belknap et al., 2020; Lee et al., 2020).

In this Chapter, we aimed to illustrate the intricate regulatory factors implicated in transduce environmental signals into metabolic outputs and to mention novel discoveries by choosing as case in point stress induced by oxidants and to translation-blocking antibiotics in *Streptomyces*.

The Genus *Streptomyces*

The bacterial genus *Streptomyces* in the largest within Actinobacteria, with approximately 900 species described to date (Lee et al., 2020). This genus is the most abundant source of naturally produced antibiotics and related compounds (Liu et al., 2013). Generally, each species produces several bioactive compounds in a species-specific way, instead of antibacterial compounds (i.e., streptomycin, vancomycin, erythromycin, and tetracycline), as well as important antifungal (amphotericin B), anticancer (mitomycin C), antiparasitic (ivermectin), and immunosuppressive (rapamycin) agents (see references in Nett et al., 2009). Other specialized metabolites include pigments, siderophores, polyketides, and peptide compounds. Bioactive compounds from *Streptomyces* are extremely important for the pharmaceutical industry (Omura et al., 2001; Bentley et al., 2002; Nett et al., 2009).

Streptomyces are distinguished by their complex life cycle (Figure 1). Vegetative growth of streptomycetes is represented by many erected hyphae on mycelial colonies. When the availability of nutrients is low, streptomycetes undergo a complex morphological process that implicates the reduction of vegetative mycelium and concludes with the formation of multiple haploid spores as a result of hyphae septation. This process occurs in parallel with the production of biologically active natural products, a process frequently called physiological differentiation, because it takes place after the main stage of active vegetative growth and assimilative metabolism (i.e., idiophase). Thus, the genus owns intricate regulatory systems that synchronize different biological processes (e.g., morphological differentiation is conducted along with physiological differentiation) and allow microorganisms to respond to environmental stresses to adapt and survive (Sevcikova et al., 2010; Traxler and Kolter, 2015; Sun et al., 2017).

Early molecular analyses revealed that *Streptomyces* possesses a long linear genome (8-10 Mbp) (Bentley et al., 2002; Nett et al., 2009). Recent advances in genomics-based approaches revealed that each *Streptomyces* species has the genetic possibility to synthesize an average of >30 natural products (Lee et al., 2020). Genome mining approaches, which frequently imply the identification of secondary metabolite biosynthetic genes, revealed that they are clustered into smBGCs that are large (usually tens of kilobases) and frequently include several operons (Liu et al., 2013; Hwang et al., 2019; Belknap et al., 2020). Next-generation sequencing (NGS) techniques and computational tools showed that *Streptomyces* genomes carry 25 to 70 smBGCs, significantly more than other actinobacterial genera (Table 1).

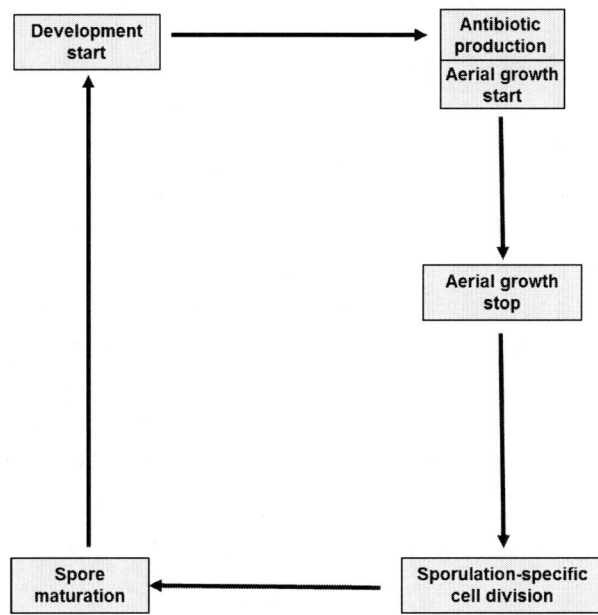

Figure 1. Major events during the life cycle of *Streptomyces* (simplified from Barka et al., 2016).

Table 1. Major classes of Biosynthetic Gene Clusters (BGCs) found in 1107 *Streptomyces* genomes (Belknap et al., 2020)

BGC	Number of genomes
Non-ribosomal peptide synthetases*	1062
Type 1 polyketide synthases (PKS)*	981
Terpenes*	697
Other ketide synthases*	650
Lantipeptides*	540
Butyrolactone	503
Type 2 PKS	499
Bacteriocin	419
Type 3 PKS	366

* These five classes of BGCs comprised approximately half of the total BGCs found in any single genome.

However, most of they remain silent under laboratory growth conditions, suggesting that the potential of streptomycetes to synthesize natural products has been under-explored and that the genus remains a valuable resource for the discovery of novel and potentially relevant pharmaceuticals, which can alleviate some of the challenges in the current antibiotic development, as mentioned before. Notwithstanding, activation of the silent smBGCs is constrained by the scarcity of regulatory information regarding their expression. To develop a deep understanding of the genetic regulatory network that targets their metabolism, it would be necessary to know the heterologous biosynthesis of smBGCs in expressing hosts (Jeong et al., 2016; Hwang et al., 2019).

MOLECULAR ADAPTATION TO THE ENVIRONMENT BY *STREPTOMYCES*

In prokaryotes, gene expression can be regulated at different levels being transcription the crucial point. At the transcriptional level, bacterial organisms regulate the expression of their genes by means of an RNA polymerase holoenzyme (RNAp), a molecule resultant from the combinatorial union of a dissociable sigma factor (σ) to an RNA polymerase core enzyme ($\alpha 2\beta\beta'\omega$). The RNAp holoenzyme is capable of using ribonucleotides to synthesize an RNA molecule (mRNA) from a specific DNA sequence that eventually can be translated into a protein. The recognition of a precise DNA template depends on the effectiveness of the sigma factor on recognizing and leading the holoenzyme to the promoter sequence of a gene (Boor, 2006; Feklístov et al., 2014; Paget, 2015; Sun et al., 2017). Since sigma factors are involved in the separation of double-stranded DNA, they play a critical role in the initiation of the transcription process; besides, once they are detached from the RNA polymerase, they can be reused (Boor, 2006).

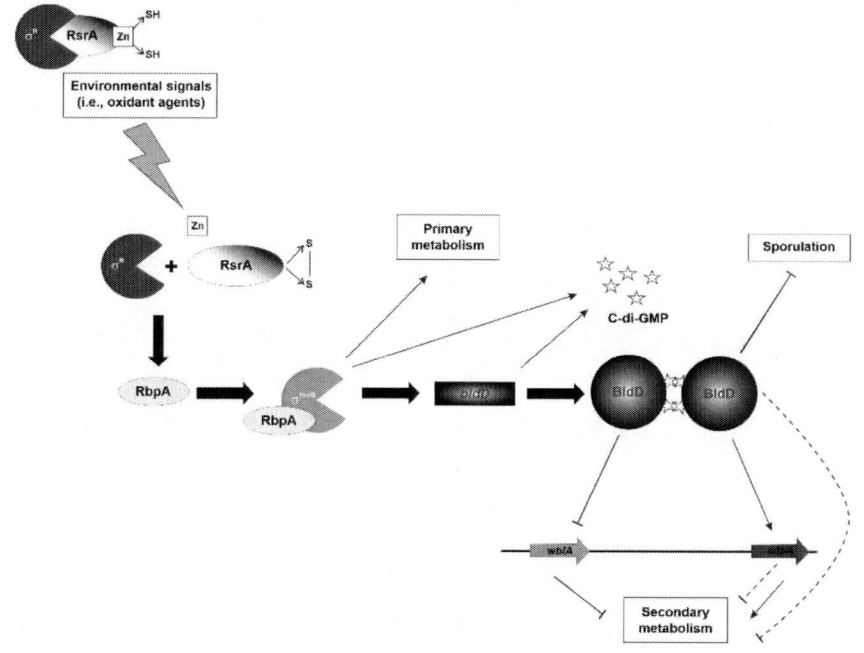

Figure 2. A general scheme for signal transduction system in *Streptomyces*. Note that the effect of BldD and AdpA on secondary metabolism is variable, depending on the species of *Streptomyces* (they act as activators in some cases, and as inhibitors and in others).

Sigma factors are not only a distinct subunit of RNAps, but also a fundamental element in the signal transduction system: their expression and activity are regulated by diverse endogenous and exogenous stimuli (Figure 2). Also, various σ factors act as master regulators since a single sigma factor can simultaneously control the expression of a large set (i.e., hundreds) of bacterial genes regulated as a union (termed regulon), thereby changing the transcriptional outcomes (Sun et al., 2017). Sigma factors are classified into two main families, the σ^{54} and the σ^{70} families. Unlike family σ^{54}, the σ^{70} family is abundant and diverse; they include a group known as 'housekeeping' sigma factors that manage the transcription of genes involved in microbial growth and metabolism (Boor, 2006; Feklístov et al., 2014; Paget, 2015; Sun et al., 2017).

The σ^{70} family is named after the *Escherichia coli* housekeeping σ^{70}, they are classified in four subgroups (Groups I, II, III, and IV) based on differential physical functions and phylogenetic relationships instead of structurally conserved domains, σ_1 (region 1.1), σ_2 (region 1.2, 2.1-2.4), σ_3 (region 3.0-3.2), and σ_4 (region 4.1-4.2). These domains are connected by flexible linkers, although some group 1 σ factors additionally contain a non-conserved region (NCR) between regions 1.2 and 2.1 (reviewed in Feklístov et al., 2014; Paget, 2015; Sun et al., 2017). Diverse morphological and physiological complexities are controlled by a proportional number of sigma factors. Thus, the σ^{70} family includes multiple members, particularly those in groups III and IV, and is highly variable among microbial species. *Streptomyces* spp. encode many σ factors (about 66 in *Streptomyces coelicolor*) (Sun et al., 2017; Šmídová et al., 2019).

When integrated to the RNA core, the DNA binding determinants are exposed, permitting promoter region recognition by the interaction of the sigma factor with promoter elements centered at -35 bp and -10 bp upstream from the transcription start site (+1) (Feklístov et al., 2014; Paget, 2015; Sun et al., 2017). Recognition of promoter sequences in response to diverse environmental signals varies between the housekeeping σ factor (σ^{HrdB}) and a variable number of alternative extracytoplasmic function (ECF) sigma factors (Group IV, conserved across both Gram-positive and Gram-negative species). These last-mentioned factors are also identified as 'general stress-responsive regulators' since they enable bacteria to cope with stressful environmental conditions by regulating the expression of genes that allow them to adapt to extracellular signals ensuring their survival (i.e., σ^R) (Boor, 2006; Staroń et al, 2009; Feklístov et al., 2014; Paget, 2015; Sun et al., 2017).

Previous observations reinforce the idea of an imminent crosstalk between ECF σ factors and housekeeping σ factors during stress conditions. Some examples are the need of σ^R to keep the proper activity of σ^{HrdB} during thiol-oxidative stress (Sun et al., 2017), enhancement of the affinity of core RNAp to main housekeeping sigma factors mediated by the transcriptional regulator RbpA (RNA polymerase binding protein A, encoded in σ^R regulon) (Tabib-Salazar et al., 2013), and that σ^{HrdB} regulon regulates the expression of a global regulator *bldD* gene (an on-off switch involved in the transition from primary to secondary metabolism) (Šmídová et al., 2019).

Extracytoplasmic Function (ECF) Sigma Factor

Sensing of the extracellular environment is achieved via a signal transduction mechanism: in the absence of a stimulus, ECF σ factors remain inactive through tightly binding to a cognate inner membrane-bound anti-σ-factor (Boor, 2006; Staroń et al, 2009). When an appropriate stimulus is perceived (i.e., redox homeostasis, cell wall integrity), the anti-σ-factor is inactivated by conformational changes or by degradation. As a consequence, the ECF σ factor is released and activated, and then it can regulate the expression of genes possessing its target promoters after recruitment by the RNA polymerase core enzyme (Staroń et al, 2009; Jiang et al, 2011).

In *Streptomyces coelicolor*, $σ^R$ (SigR), a member of ECF sigma factors of Group IV, is induced by oxidants and translation-blocking antibiotics. This alternative sigma factor was named SigR to make reference to redox regulation (Sun et al., 2017; Park et al., 2019).

The *sigR* gene is positively autoregulated since its anti-sigma factor, RsrA, is part of the $σ^R$ regulon. RsrA contains the HX_3CX_2C motif that is characteristic of the zinc-containing anti-sigma factor family (ZAS). This family groups genes usually located downstream of those coding for ECF σ factors (Kang et al., 1999; Staroń et al., 2009). Under reducing conditions, $σ^R$ remain inactive by binding with RsrA. Simultaneously with the occurrence of oxidant compounds (i.e., hydrogen peroxide or diamide) and thiol-reactive electrophiles, disulfide bond formation or alkylation of reactive cysteines occurs on RsrA, any of which results in inactivation of the anti-sigma factor and further release of $σ^R$, thus, allowing it to bind to the core RNAp enzyme (Kang et al., 1999). Transcriptional activation of *sigR* promoter via antibiotic induction is mediated by WblC, a WhiB-like regulator that confers resistance to antibiotics in streptomycetes and mycobacteria (Park et al., 2019). The expression of $σ^R$ is regulated at multiple molecular levels: at the transcriptional level, it is promoted by WblC, at the translational level it is repressed by translation initiation factor IF3 (which is regulated by σ^{HrdB}, as seen below), and at the post-transcriptional level by Clp proteases (Sun et al., 2017).

Transcriptome and ChIP analyses revealed principal functional categories controlled by the SigR regulon, including thiol homeostasis, sulfur metabolism, ribosome modulation, guanine nucleotide metabolism, protein degradation, energy metabolism, lipid metabolism, DNA repair and recombination, cofactor metabolism, and transcriptional regulators (i.e., SigR-RsrA, RsrA2-SigR2-RsrA2-1, NdgR, HrdB, HrdD, and RpbA) (Park et al., 2019).

Housekeeping σ^{HrdB} Factor

Previous evidence has pointed out that instead of being required to control gene expression, the housekeeping σ^{HrdB} is involved in morphological and physiological differentiation; thus, playing an important role in redirecting the metabolic flux from primary to secondary metabolism. Not surprisingly, molecular engineering of the *hrdB* gene has been employed as a strategy to enhance natural compounds biosynthesis (i.e., antibiotics) (Sun et al., 2017).

Because previous efforts to explore the natural function of the HrdB regulon by deleting *hrdB* was fatal for bacteria, Šmídová et al. (2019) constructed a strain carrying a HrdB protein with the hemagglutinin (HA) tag, which allowed them to determine the promoter binding process by ChIP-Seq. Also, *in silico* modeling of gene expression kinetics was used with either the HrdB or the Hrd-RbpA complex as regulators. A total of 2137 genes were found by ChIP-seq, of which 1694 were controlled either by HrdB or Hrd-RbpA complex; gene class abundance among genes was settled into specific categories summarized in Tables 2 and 3.

Table 2. Main functional groups regulated by HrdB or HrdB-RbpA complex in *S. coelicolor* (Šmídová et al., 2019)

Functional group	Total of genes	Genes under control of HrdB or HrdB-RbpA complex	% regulated	Mostly integrated by
Chomosome replication	8	5	62.5	DNA polymerase subunits
Nucleotide biosynthesis	30	23	76.6	purine/pyrimidine metabolism
Ribosome constituents	67	47	70.1	ribosomal proteins

Table 3. Sigma factors and sigma-related factors regulated by HrdB or HrdB-RbpA complex in *S. coelicolor* (Šmídová et al., 2019)

Factors	Genes under control of HrdB or HrdB-RbpA complex
Sigma factors	1723 (sigK), 3202 (hrdD), 5621 (whiG), 0037, 0038, 0414, 2639, 3709, 4336, 4409, 6996, 7278
Anti-sigma factors	0599 (rsbA)
Anti-sigma factor antagonists	3549(bldG), 0869, 3692, 4027
Anti-anti-sigma factor	3067

Also, they identified an important number of genes whose expression is controlled by HrdB (genes into the HrdB regulon). Interestingly, these genes participate in key processes in the cell: glucose, fatty acid, nucleotide and ribosome biosynthesis, energy and amino acid metabolism, genome replication, transcription and translation processes, as well as cell division. Besides, *bld* genes (i.e., *bldD*, *bldN*, and *bldg*) were identified, some of which are involved in morphological differentiation and antibiotic biosynthesis (i.e., *bldD*) (den Hengst et al., 2010; Šmídová et al., 2019).

On the other hand, the use of transcriptional factor RbpA as an additional regulator (HrdB-RbpA) revealed that the expression profiles of only 322 of the 1694 genes controlled by HrdB were improved; they are implicated mainly in genome replication, adaptation, energy metabolism, nucleotide, and ribosome biosynthesis. Additionally, they identified 41 genes whose expression could be exclusively regulated by the HrdB-RbpA complex. Unfortunately, they were unable to identify any functional group in particular. Curiously, ontological analysis of genes found not to be under the control or either HrdB or Hrd-RbpA revealed that they belong to 'secondary metabolism' (30 of a total of 277 genes), particularly from the polyketide biosynthetic pathway (Šmídová et al., 2019).

Transcriptional Regulator, BldD

As previously mentioned, in streptomycetes the beginning of morphological differentiation occurs simultaneously with the production of secondary metabolism (physiological differentiation). The progression of morphological differentiation requires the activity of two main classes of regulatory genes, white (*whi*) and bald (*bld*) as are called the mutants in these genes. *whi* mutants develop aerial hyphae in a normal way; however, these hyphae are incapable of completing the differentiation cycle to form mature chains of spores. Contrarily, *bld* mutants are unable to form aerial hyphae and thus appear 'bald'. It is common that mutants with either *bld* or *whi* mutations are also deficient in secondary metabolite production (den Hengst et al., 2010; Li et al., 2019; Makitrynskyy et al., 2020; Yan et al., 2020).

Some studies have reported that BldD is essential for the normal developmental program in microorganisms because deficient BldD mutants in *Streptomyces* spp. were blocked at the early stages of the growth cycle. Also, BldD is fundamental in the biosynthesis of actinorhodin, erythromycin, prodigionines, methylenomycin, chloramphenicol, and calcium-dependent antibiotic (Cda) (Li et al., 2019; Yan et al., 2020).

BldD is widespread among sporulating actinomycetes, generally functioning as a repressor of genes required for morphological differentiation and antibiotic production. It controls more than 167 regulons in *S. coelicolor*. The consensus BkdD-binding site 'BldD box' is a 15 bp palindromic sequence 5-NTNACNC(A7t)GNGTNAN-3 (den Hengst et al., 2010).

However, recent works have demonstrated BldD's role as a transcriptional activator by directly activating the biosynthesis of lincosamide antibiotics, i.e., lincomycin by *Streptomyces lincolnensis* by BldD activating transcription of lincomycin biosynthetic structural genes or lincomycin CSR *lmbU* (Li et al., 2019). Yan et al. (2020) reported that BldD directly activated the production of the anthelmintic agent, avermectin, by *Streptomyces avermitilis* (i.e., by directly activating *ave* gene cluster or activating the transcription of CSR gene *aveR*). In the case of biosynthesis of cyclic lipopeptide antibiotic, daptomycin (dpt cluster) by *Streptomyces roseosporus*, they reported that BldD could straightly activate daptomycin production by directly switching-on the transcription of *dpt* operon or via a cascade mechanism by starting the transcription of pleiotropic regulatory genes (i.e., *afsR* and *adpA)*. Similarly, Makitrynskyy et al., (2020) reported the positive regulation of AdpA by BldD in *Streptomyces ghanaensis*, leading to the production of the pentasaccharide antibiotic, moenomycin A (MmA).

BldD regulon genes include development-related genes (i.e., *bldA*, *bldC*, *bldD*, *bldN*, *bldM*, *sigH*, *whiB*, *whiG*, among others), which have been implicated in antibiotic and polysaccharide production. Besides, the AdpA (also called BldH) regulon and the diguanylate cyclase gene *cdgA* (cyclic dimeric 3'-5' guanosine monophosphate, c-di-GMP) has been implied as BldD targets (den Hengst et al., 2010). Later, Tschowri et al. (2014) discovered that the regulatory activity of BldD is controlled post-translationally by the cyclic-di-GMP (c-di-GMP) signaling molecule. In presence of c-di-GMP, the BldD C-terminal domain identifies a c-di-GMP tetramer bipartite signature 'RXD-X_8-RXXD' that mediates the production of a dimeric BldD, which is, then, able to bind to its target genes.

In bacteria, c-di-GMP is synthesized by the condensation of two molecules of GTP, a process catalyzed by guanylate cyclases (DGCs) (interestingly, a **diguanylate cyclase gene cdgB**, belong to σ^{HrdB} regulon). Degradation can be achieved in two different ways by phosphodiesterases (PDEs) (Latoscha et al., 2019; Sivapragasam and Grove, 2019;

Makitrynskyy et al., 2020). In *Streptomyces*, the signaling network of c-di-GMP comprises more than 11 metabolic enzymes, some of them related to their own synthesis (CdgA, CdgB, CdgC, and CdgD) and degradation (RmdA and RmdB) (den Hengst et al., 2010; Latoscha et al., 2019).

In microbial organisms, c-di-GMP along with other important nucleotide-based second messengers (i.e., guanosine tetra- and pentaphosphates (collectively termed as '(p)ppGpp', also called alarmone, cAMP and c-di-AMP) are fundamental for transducing signals in response to stress and starvation (i.e., to generate molecular mechanisms to survive harmful environmental circumstances). In streptomycetes, these signaling molecules control diverse biological processes such as morphological differentiation and biosynthesis of natural products, including antibiotics. Under stressful conditions, they require the presence of purines to maintain the synthesis of these guanosine products (Latoscha et al., 2019; Sivapragasam and Grove, 2019).

As mentioned above, recently, Makitrynskyy et al. (2020) and Yan et al. (2020) reported that, in *S. ghanaensis* and *S. roseosporus*, respectively, the transcription of the *adpA* gene (in *Streptomyces*, a global regulator of morphological differentiation and secondary metabolite biosynthetic gene clusters) was directly regulated by BldD in a positive way, leading to the production of the moenomycin A and daptomycin antibiotics, respectively. Those results are opposite to those previously reported for *S. coelicolor*, where BldD represses the transcription of *adpA* (den Hengst et al., 2010). Newly, a correlation between expression levels of *adpA* and intracellular concentration of c-di-GMP has been described (Makitrynskyy et al., 2020).

Based on molecular and biochemical data, Makitrynskyy et al. (2020) found that, in streptomycetes, the 'bldD-c-di-GTP' complex inhibits spore formation (i.e., transcriptionally repressing regulatory gene *wblA*, (generally involved in late stages of morphological differentiation), and allows continuous biosynthesis of natural products (i.e., directly activating *adpA*). Also, importantly, they found that c-di-GMP regulates expression levels of *bldD*, and that BldD interacts with the *cdgB* promoter producing a reciprocal regulatory loop in *Streptomyces ghanaensis*. Deletion of an active PDE in this bacterium (RmdB) led to a massive increase in secondary metabolites production, including the macrolide antibiotic oxohygrolidin and the clinically important desferrioxamine. Molecular engineering of genes involved in c-di-GMP synthesis could improve antibiotic production, including those coded in cryptic smBGCs. Besides, detailed studies of roles of transcriptional regulators BldD and AdpA in development and secondary metabolite need to be conducted in other *Streptomyces* species.

REFERENCES

Barka, E. A., Vatsa, P., Sanchez. L., Gaveau-Vaillant, N., Jacquard, C., Klenk, H. P., Clément, C., Ouhdouch, Y., van Wezel, G. P. (2016). Taxonomy, physiology, and natural products of Actinobacteria. *Microbiology and Molecular Biology Reviews, 80*, 1-43. https://doi.org/10.1128/MMBR.00019-15.

Belknap, K. C. Park, C. J. Barth, B. M. Andam, C. P. (2020) Genome mining of biosynthetic and chemotherapeutic gene clusters in *Streptomyces* bacteria. *Scientific Reports 10*, 2003. https://doi.org/10.1038/s41598-020-58904-9.

Bentley, S. D., Chater, K. F., Cerdeño-Tárraga, A. M., Challis, G .L., Thomson, N. R., James, K. D., Harris, D. E., Quail, M. A., Kieser, H., Harper, D., Bateman, A., Brown, S., Chandra, G., Chen, C. W., Collins, M., Cronin, A., Fraser, A., Goble, A., Hidalgo, J., Hornsby, T., Howarth, S., Huang, C. H., Kieser, T., Larke, L., Murphy, L., Oliver, K., O'Neil, S., Rabbinowitsch, E., Rajandream, M. A., Rutherford, K., Rutter, S., Seeger, K., Saunders, D., Sharp, S., Squares, R., Squares, S., Taylor, K., Warren, T., Wietzorrek, A., Woodward, J., Barrell, BG., Parkhill, J., Hopwood, D. A. (2002). Complete genome sequence of the model actinomycete *Streptomyces coelicolor* A3(2). *Nature, 417*, 141-147. https://doi.org/ 10.1038/417141a18

Bérdy, J. (2012). Thoughts and facts about antibiotics: where we are now and where we are heading. *Journal of Antibiotics 65*, 385-395. https://doi.org/10.1038/ja.2012.27.

Boor, K. J. (2006). Bacterial stress responses: what doesn't kill them can make them stronger. *PLoS Biology 4*, e23. https://doi.org/10.1371/journal.pbio.0040023.

Cornforth, D., Foster, K. (2013). Competition sensing: the social side of bacterial stress responses. *Nature Reviews Microbiology. 11*, 285-293. https://doi.org/10.1038/nrmicro2977.

den Hengst, C. D., Tran, N. T., Bibb, M. J., Chandra, G., Leskiw, B. K., Buttner, M.J. (2010). Genes essential for morphological development and antibiotic production in *Streptomyces coelicolor* are targets of BldD during vegetative growth. *Molecular Microbiology 78*, 361-379. https://doi.org/10.1111/j.1365-2958.2010.07338.x.

Feklístov. A., Sharon, B. D., Darst, S. A., Gross, C. A. (2014). Bacterial sigma factors: a historical, structural, and genomic perspective. *Annual Reviews in Microbiology 68*, 357-76. https://doi.org/10.1146/annurev-micro-092412-155737.

Hwang, S., Lee, N., Jeong, Y., Lee, Y., Kim, W., Cho, S., Palsson, B. O., Cho, B. K. (2019). Primary transcriptome and translatome analysis determines transcriptional and translational regulatory elements encoded in the *Streptomyces clavuligerus* genome. *Nucleic Acids Research 47*, 6114-6129. https://doi.org/10.1093/nar/gkz471.

Jeong, Y., Kim, J., Kim, M., Bucca, G., Cho, S., Yoon, Y. J., Kim, B. G., Roe, J. H., Kim, S. C., Smith, C. P., Cho B. K. (2016). The dynamic transcriptional and translational landscape of the model antibiotic producer *Streptomyces coelicolor* A3(2). *Nature Communications 7*, 11605. https://doi.org/10.1038/ncomms11605.

Jiang, L., Liu, Y., Wang, P., Wen, Y., Song, Y., Chen, Z., Li. J. (2011). Inactivation of the extracytoplasmic function sigma factor Sig6 stimulates avermectin production in *Streptomyces avermitilis*. *Biotechnology Letters 33*, 1955-1961. https://doi.org/10.1007/s10529-011-407 0673-x.

Kang, J. G., Paget, M. S., Seok, Y. J., Hahn, M. Y., Bae, J. B., Hahn, J. S., Kleanthous, C., Buttner, M. J., Roe, J. H. (1999). RsrA, an anti-sigma factor regulated by redox change. *EMBO Journal 18*, 4292-4298. https://doi.org/10.1093/emboj/18.15.4292.

Latoscha, A., Wörmann, M. E., Tschowri, N. (2019). Nucleotide second messengers in *Streptomyces*. *Microbiology 165*, 1153-1165. https://doi.org/10.1099/mic.0.000846.

Lee, N., Kim, W., Hwang, S., Lee, Y., Cho, S., Palsson, B., Cho, B.K. (2020). Thirty complete *Streptomyces* genome sequences for mining novel secondary metabolite biosynthetic gene clusters. *Scientific Data 7*, 55. https://doi.org/10.1038/s41597-020-0395-9.

Li, J., Wang, N., Tang, Y., Cai, X., Xu, Y., Liu, R., Wu, H., Zhang, B. (2019). Developmental regulator BldD directly regulates lincomycin biosynthesis in *Streptomyces lincolnensis*. *Biochemical and Biophysical Research Communications 518*, 548-553. https://doi.org/10.1016/j.bbrc.2019.08.079.

Liu, G., Chater, K. F., Chandra, G., Niu, G., Tan, H. (2013). Molecular regulation of antibiotic biosynthesis in *Streptomyces*. *Microbiology and Molecular Biology Reviews 77*, 112-143; https://doi.org/10.1128/MMBR.00054-12.

Makitrynskyy, R., Tsypik, O., Nuzzo, D., Paululat, T., Zechel, DL., Bechthold, A. (2020). Secondary nucleotide messenger c-di-GMP exerts a global control on natural product biosynthesis in *Streptomycetes*. *Nucleic Acids Research 48*, 1583-1598. https://doi.org/10.1093/nar/gkz1220.

Nett, M., Ikeda, H., Moore, B. S. (2009). Genomic basis for natural product biosynthetic diversity in the actinomycetes. *Natural Product Reports 26*, 1362-1384. https://doi.org/10.1039/b817069j.

Omura, S., Ikeda, H., Ishikawa, J., Hanamoto, A., Takahashi, C., Shinose, M., Takahashi, Y., Horikawa, H., Nakazawa, H., Osonoe, T., Kikuchi, H., Shiba, T., Sakaki, Y., Hattori, M. (2001). Genome sequence of an industrial microorganism *Streptomyces avermitilis*: deducing the ability of producing secondary metabolites. *Proceedings of the National Academy of Sciences of the United States of America 98*, 12215-12220. https://doi.org/10.1073/pnas.211433198.

Paget, M. S. (2015). Bacterial sigma factors and anti-sigma factors: structure, function and distribution. *Biomolecules 5*, 1245-1265. https://doi.org/10.3390/biom5031245.

Park, J. H., Lee, J. H., Roe, J. H. (2019). SigR, a hub of multilayered regulation of redox and antibiotic stress responses. *Molecular Microbiology 112*, 420-431. https://doi.org/10.1111/mmi.14341

Pham, J. V., Yilma, M. A., Feliz, A., Majid, M. T., Maffetone, N., Walker, J. R., Kim, E., Cho, H. J., Reynolds, J. M., Song, M. C., Park, S. R., Yoon, Y. J. (2019). A review of the microbial production of bioactive natural products and biologics. *Frontiers in Microbiology 10*, 1404. https://doi.org/10.3389/fmicb.2019.01404.

Sevcikova. B., Rezuchova, B., Homerova, D., Kormanec, J. (2010). The anti-anti-sigma factor BldG is involved in activation of the stress response sigma factor σH in *Streptomyces coelicolor* A3(2). *Journal of Bacteriology 192*, 5674-5681. https://doi.org/10.1128/JB.00828-10.

Sivapragasam, S., Grove, A. (2019). The link between purine metabolism and production of antibiotics in *Streptomyces*. *Antibiotics 8*, 76-88. https://doi.org/10.3390/antibiotics8020076.

Šmídová, K., Ziková, A., Pospíšil, J., Schwarz, M., Bobek, J., Vohradsky, J. (2019). DNA mapping and kinetic modeling of the HrdB regulon in *Streptomyces coelicolor*. *Nucleic Acids Research 47*, 621-633. https://doi.org/10.1093/nar/gky1018.

Staroń, A., Sofia, H. J., Dietrich, S., Ulrich, L. E., Liesegang, H., Mascher, T. (2009). The third pillar of bacterial signal transduction: classification of the extracytoplasmic function (ECF) σ factor protein family. *Molecular Microbiology 74*, 557-581. https://doi.org/10.1111/j.1365- 2958.2009.06870.

Sun, D., Liu, C., Zhu, J., Liu, W. (2017). Connecting metabolic pathways: sigma factors in *Streptomyces* spp. *Frontiers in Microbiology 8*, 2546. https://doi.org/10.3389/fmicb.2017.02546.

Tabib-Salazar, A., Liu, B., Doughty, P., Lewis, R. A., Ghosh, S., Parsy, M. L., Simpson, P. J., O'Dwyer, K., Matthews, SJ., Paget, M. S. (2013). The actinobacterial transcription factor RbpA binds to the principal sigma subunit of RNA polymerase. *Nucleic Acids Research 41*, 5679-5691. https://doi.org/457 10.1093/nar/gkt277.

Traxler, M. F., Kolter, R. (2015). Natural products in soil microbe interactions and evolution. *Natural Product Reports 32*, 956-970. https://doi.org/10.1039/c5np00013k.

Tschowri, N., Schumacher, M. A., Schlimpert, S., Chinnam, N. B., Findlay, K. C., Brennan, R. G., Buttner, M. J. (2014). Tetrameric c-di-GMP mediates effective transcription factor dimerization to control *Streptomyces* development. *Cell 158*, 1136-1147. https://doi.org/10.1016/j.cell.2014.07.022.

van der Heul, H. U., Bilyk, B. L., McDowall, K. J., Seipke, R. F., van Wezel, G. P. (2018). Regulation of antibiotic production in Actinobacteria: new perspectives from the post-genomic era. *Natural Product Reports 35*, 575-604. https://doi.org/10.1039/c8np00012c.

Wei, J., He, L., Niu, G. (2018). Regulation of antibiotic biosynthesis in actinomycetes: Perspectives and challenges. *Synthetic and Systems Biotechnology 3*, 229-235. ttps://doi.org/10.1016/j.synbio.2018.10.005.

Yan, H., Lu, X., Sun, D., Zhuang, S., Chen, Q., Chen, Z., Li, J., Wen, Y. (2020). BldD, a master developmental repressor, activates antibiotic production in two *Streptomyces* species. *Molecular Microbiology 113*, 123-142. https://doi.org/10.1111/mmi.14405.

In: Molecular Basis of Specific Mechanism ... ISBN: 978-1-53618-751-9
Editors: Marcos López-Pérez et al. © 2020 Nova Science Publishers, Inc.

Chapter 4

MOLECULAR BASIS OF BACTERIAL HOMEOSTASIS UNDER ENVIRONMENTAL STRESS AND CELLULAR TRANSPORT AT MEMBRANE LEVEL

Marcos López-Pérez[*] *and Félix Aguirre Garrido*
Environmental Sciences Department, Metropolitan Autonomous University (Lerma Unit) Lerma de Villada, México

ABSTRACT

From the perspective of the dynamics of the microbial ecological niche, abrupt environmental variation has been constant in the history of the biosphere. The most relevant thing in this context is to achieve adaptation as quickly as possible; in this sense, the process involves activated mechanisms, which, in cellular physiology, converge and can be broadly summarized as responsible for the maintenance of homeostatic balance. This chapter compiles the molecular bases of the most representative mechanisms occurring in the plasma membrane, a structure

---

[*] Corresponding Author's Email: m.lopez@correo.ler.uam.mx.

that regulates transport between the cell and the surrounding environment, a core element for homeostatic regulation.

Keywords: bacterial homeostasis, stress, membrane

1. Introduction: General Information on Transportation at Membrane Level and Homeostasis

Membranes have a central and fundamental role in cell physiology. From the structural point of view, they not only represent a barrier that separates the cytoplasm from the external environment, acting passively by regulating the passage of certain elements (Gatenby, 2019), but they also act as a support for cellular signaling (Sunshine & Iruela-Arispe 2017), or the recruitment of proteins (Falke, 2007); besides, they are directly related to the transport regulation. Structurally, the membrane is an intrinsically plastic structure capable of maintaining its activity under different environmental disturbances (Zhou & Jin 2018). One element that is pertinent to mention refers to studies carried out decades ago that implicated adaptation phenomena associated to membranes through composition changes as a consequence of temperature variation (Sinensky 1974; Sinensky 1980; Cronan & Gelmann 1975). In this same sense, it is possible to cite works that allude to specific characteristics of the membranes in response to pH conditions (Angelova et al. 2018), salinity, high solute concentrations (Wood 2015), or even abiotic variables such as intense radiation (Sherif et al. 2011). Throughout this chapter, the molecular bases of the action mechanisms associated with the plasma membrane involved in the homeostatic maintenance regulation, whose ultimate purpose is to achieve adequacy, will be explored.

2. PHYSICAL-CHEMICAL MECHANISMS OF PERCEPTION AND ENVIRONMENTAL RESPONSE IN THE MEMBRANE FOR HOMEOSTATIC REGULATION

From a classical perspective, cells have been shown to be able to adapt their composition and physical properties of their membrane in response to environmental change (Sinensky 1974; Sinensky 1980). In this sense, there are different examples, such as *E. coli*, where lipid substitution with a different unsaturation degree has been reported to optimally maintain membrane viscosity (Sinensky 1974). This type of adaptation is called homeoviscous, which is shown in Figure 1, and has been reported in other prokaryotes and even other organisms, such as ectothermic chordates (Ernst 2016; Hazel 1995). The molecular bases of these processes are relevant for *Bacillus subtillis*, where the DesK/DesR genes related to the thickness regulation of the lipid bilayer have been studied in depth, which, in turn, is directly related to membrane viscosity (Saita et al. 2016; Cybulski et al. 2010).

Another parameter that must be considered to address the physical-chemical mechanisms refers to the abundance and composition of integral membrane proteins (Ballweg and Ernst 2017). In this sense, it is interesting to emphasize that it is much faster to remodel the lipidome of the membranes than its proteome, therefore, it is reasonable to infer that changes related to lipids are the first response to environmental disturbances (Tulodziecka et al. 2016).

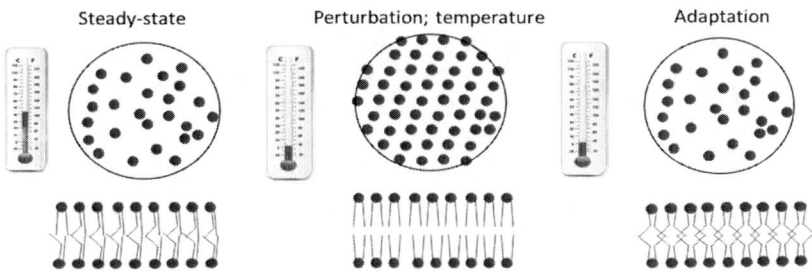

Figure 1. Homeoviscous mechanism, image modified from Ernst et al. 2017.

Table 1. Most relevant mechanisms related with membrane properties and homeostatic regulation

Membrane property	Major determinant	Potential sensing mechanisms
Lateral pressure in membrane core	Lipid headgroups and acyl chains, lipid shape	Gradual folding of an AH in packing defects/CCT (Cornell 2016), Squalene monooxygenase (Chua et al. 2017)
Membrane compressibility	Lipid headgroup interactions, lipid packing, sterol content	Local membrane compression stabilizes oligomeric state when
Lipid packing/membrane order	Lipid saturation, sterol content	Mga2 (Covino, et al. 2016 Ballweg & Ernst, 2017), CCT (Cornell, 2016), squalene monooxygenase (O'Hara et al., 2006)
Surface charge density	Concentration of charged lipids	Sensing of electrostatics by amphipathic helix with basic residues/CCT (Cornell 2016), Pah1 (Saita et al. 2016)
Membrane thickness	Hydrophobic thickness of proteins, length of lipids, sterol content	Hydrophilic residues drive conformational changes when thickness increases/DesK (Cybulski, 2010; Groisman et al, 2013)
Intrinsic curvature	Lipid shape	Sensing at surfaces and across the bilayer is required
Stiffness/bending rigidity	Lipid headgroups and acyl chains, lipid shape, electrostatics	Motor proteins could challenge membrane rigidity upon membrane bending and tubulation
Lateral pressure at membrane surface	Lipid headgroups and acyl chains, sterol content, lipid shape	Conformational changes (e.g., helix: helix rotations) sensitive to the lateral pressure profile/Mga2 (Covino et al. 2016; Ballweg & Ernst 2017)
Membrane fluidity/viscosity		Sensing the frequency of collisions of membrane constituents

Table obtained from Ernst et al., 2017.

From a qualitative perspective, another interesting element that should be mentioned regarding modifications in lipid membranes, is limited to which lipids are replaced in the membranes and, therefore, which metabolic pathways in the bacterium's cellular physiology are affected (Zhang and Rock 2008). Hence, membrane properties are largely defined by the fatty acid structures that compose them. Table 2 summarizes some of the main phospholipids present in bacterial membranes.

Table 2. Principals' phospholipids present in bacterial membranes

Structure	Fatty acid	Effect
	Trans-11-C18:1	Mimics saturated fatty acids, provides resistance to solvents,
	Cis 11-C18:1	Increases membrane fluidity
	Iso-C17:0	Decreases membrane fluidity compared with anteiso-chains
	Anteiso-C17:0	Increases membrane fluidity compared with iso-chains
	Cyclopropane-C17O	Mimics unsaturated fatty acids and increases stability to acid stress
	C16:0	Decreases membrane fluidity

3. ROLE OF EFFLUX SYSTEMS ASSOCIATED WITH STRESS ATTENUATION BY THE PRESENCE OF HEAVY METALS

Mechanisms related to cell efflux have traditionally been associated with the resistance to drugs with antimicrobial activity, mainly due to the clinical and social relevance that this implies (Davies & Davies 2010). On the other hand, it has been published in several works that these mechanisms are not exclusive to bacterial cell physiology, but are present in a large number of organisms, in which genes with a high conservation degree in natural history have been determined that are finely regulated. These mechanisms are associated with different functions in bacterial physiology, such as virulence in pathogenic processes (Li & Nikaido 2009), detoxification of secondary metabolites (Blanco et al. 2016), transport phenomena that are implicitly related to the purpose of this heading, and maintenance of homeostatic conditions necessary for cellular processes (Martinez et al, 2009). The presence of heavy metals has been constant in bacterial ecosystems, since they are part of minerals present in the soil, which through different mechanisms dissolve in water and, consequently, affect organisms (Singh et al. 2011). Throughout evolutionary history, bacterial physiology has developed a great diversity of mechanisms that have enabled bacteria to adapt to the cited conditions, which can be broadly categorized into two large groups. 1) Non-specific, impermeability or detoxifications by chelation; for example, the complex formation or chelation with bacterial extracellular polymers (Krul 1977). On the other hand, Brandt et al. (2006) studied in depth the growth physiology of Pseudomonas, and found that they produce significant amounts of extracellular polymers in the presence of copper. In reference to this large group of non-specific strategies, it is pertinent to indicate that although they allow adaptation to the presence of any type of heavy metal, it has also been reported that the concentrations at which they allow this adaptation are much lower when compared to the other large group of mechanisms, which are called specific. In particular, it has been reported that specific resistance mechanisms provide bacteria that carry them a resistance thousands of times higher than non-specific ones (Guha et al. 1981). In reference to these two categories of mechanisms, it can be said that non-specific mechanisms are characterized by being related to very general aspects of growth physiology, such as cell aggregation, which implies the identification of a critical cell density, widely described in the literature, as is the case of the quorum sensing system (Miller and Bonnie 2001). In general terms, it is possible to refer to a resistance phenotype, which involves general aspects of growth physiology that have two basic goals. On the one hand, to decrease the area-volume ratio, thereby allowing the cells a degree of minor exposure to a stressful agent and, therefore, buying time to achieve adequacy. On the other hand, reducing the distance between cells, potentially allowing for the exchange of genetic information through horizontal gene transfer (HGT) processes, which increases the likelihood of adequacy since it allows bacteria to maximize acquired resistance. Regarding the specific mechanisms, there are multiple works that have explored this aspect of microbial adaptation. For example, studies performed with *Staphylococcus aureus* carrying penicillinases that have been related to resistance to the presence of arsenate, arsenite, cadmium, mercury, and bismuth (Götz et al. 2002). One of the systems that are active in both non-specific and specific mechanisms are proteins related to cell efflux. In this regard, there are very well-articulated review works, such as the work by Nies in 2003, who made a very detailed review of the mechanisms associated with resistance to the presence of heavy metals. Other studies have explored MDR (multidrug resistance pumps) efflux mechanisms and their relationship with the adaptation processes to the presence of heavy metals (Silver & Phung 1996, 2005). Part of the reasons that cell efflux mechanisms are common for antibiotics, toxins, and heavy metals is that all of these elements are a natural part of ecosystems and, therefore, have been in contact with bacteria since the origin of life (Silver & Phung 2005). In this sense, it is to be expected that the unspecific adaptation mechanisms are related to the function of expelling elements from the cytoplasm that generate any form of stress. However, there are also proteins related to the cellular flow associated with the expulsion of heavy metals that are highly specific (Saier et al. 1998). Some of the reasons that explain the relationship between the heavy metal cell efflux associated with homeostasis is that many enzymes require heavy metals in trace concentrations as cofactors and, therefore, the intracellular concentration of these metals must be regulated finely to ensure the proper operation. In this sense, there are several works, for example in *Pseudomonas aeruginosa,* that have demonstrated the relationship between the expressions of proteins associated with cell efflux and the presence of heavy metals (Perron et al. 2004; Teitzel et al. 2006).

4. BACTERIAL HOMEOSTASIS AND DETOXIFICATION BASED ON MDR TRANSPORT

The detoxification processes and homeostasis maintenance are intrinsically related in cellular physiology. Detoxification acts on the cell at two levels, on the one hand at the intracellular level, where for several decades several enzymatic activities associated with detoxification have been described, such as the reductase activity associated with detoxification of mercury (Robinson & Tuovinen 1984) or the acetyl transferase activity to attenuate the effects derived from the presence of cobalt and nickel (Freeman et al., 2005). On the other hand, other mechanisms that detoxify the cell are transport-related proteins integrated into the plasma membrane, basically MDR proteins. Traditionally, these proteins have been studied based on clinical approaches, where they have been associated with the resistance to antibiotics, particularly they have been related to the removal process of this type of secondary metabolites. Although, for this specific type of transporters, this relationship is not clear yet, for example, *Streptomyces* species are typically antibiotic-producing microorganisms; however, they have been shown to carry a large number of genes that are potentially capable of conferring resistance when expressed heterologously in other organisms, at the same time, it has been shown that they are susceptible to a variety of drugs with antimicrobial activity (Lee et al., 2006, 2003). Along these same lines, several works have reported increasing evidence that suggests that the membrane-associated MDR systems are involved in the elimination of toxic compounds, generated from bacterial metabolism (Neyfakh, 1997). In this regard, for example, the *Pseudomonas auroginosa* MexGHI-OpmD MDR pump has been related to the influx of anthranilate, a toxic precursor of the *Pseudomonas* quinolone signal (Aendekerk et al. 2002, 2005; Sekiya et al. 2003). Finally, the fundamental role of MDR systems in maintaining cellular homeostasis should be highlighted, as it has been described that this type of protein contributes to homeostatic balance. Another example is the *E. coli* MdfA transporter, a system that has been related to the mechanism of Na1 (K1)/H1 antiporter that enables cells to maintain intracellular pH homeostasis under alkaline conditions (Lewinson & Bibi, 2001).

5. TRANSPORT REGULATION, HOMEOSTASIS, AND IMPLICATIONS OF VIRULENCE IN PATHOGENIC BACTERIA

One of the mechanisms that determines drastic changes in bacterial physiology involves the transition from a saprophytic or non-infectious form to a virulent phenotype based on homeostatic regulation related to transport. In this sense, the importance of Mg^{2+} concentrations in relation to the stabilization of membranes and ribosomes has been described, it has also been related to an intracellular neutralization effect of nucleic acids, as well as its role as a cofactor in a variety of enzymatic reactions (Groisman et al. 2013). One of the microorganisms in which regulation of Mg^{2+} concentration associated with homeostasis and virulence has been best studied is *Salmonella enterica*. The first works that addressed this theoretical framework identified various membrane transporters in *Salmonella* (Hmiel 1986; Hmiel 1989; Snavely 1989). On the other hand, the presence of a virulence regulator controlled by transcription factors associated with Mg^{2+} transporters has been demonstrated (Garcia Vescovi 1996). Extracellular control of magnesium concentration has been studied with special attention to genes associated with the expression and activity control of Mg^{2+} transporters. Three different classes of Mg^{2+} transporters have been described in bacteria: CorA, MgtE, and MgtA (Hmiel 1986; Hmiel 1989, Smith 1995). The most interesting operation model is based on the crystallographic analysis of the transporter PhoQ, from *Samonella*, which is based on the negative charge of some amino acids located in the periplasmic space, it seems that the relation among charges plays a fundamental role that is established at that particular point and is affected by these ions with a net positive charge, the different concentrations of the medium destabilize this part of the protein structure, leading it to transit to different degrees of activity (Cho et al. 2006; Garcia Vescovi 1996). On the other hand, the presence of cytoplasmic systems for detecting Mg concentration has also been documented (Cromie et al. 2006; Dann et al. 2007). They induce or repress the expression of transporters depending on the cytoplasmic magnesium demand. The best described mechanism in the literature that regulates the extra-cytoplasmic magnesium uptake system is based on riboswitches. Particularly metal sensing riboswitches (Cromie et al. 2006). The riboswitches, in a generic way, can be said to be fragments of mRNA that specifically recognize certain ligands and are capable of modifying their morphology and structure when they bind to the signal molecule. Under these structural changes, these mRNA molecules modify their expression patterns (Henkin 2008; Roth and Breaker 2009). One of the hypotheses that magnesium concentration is related to virulence involves the fact that the PhoP transporter, which is activated by the presence of Mg, contributes to the pathogenicity and virulence of *Salmonella,* because the mgtCBR operon is located in the SPI-3 pathogenicity island, containing the genes required to transition to a more virulent form in a pathogenesis process (Blanc-Potard 1997). This hypothesis is also supported by the fact that the MgtB protein works at 37°C but not at 20°C (Snavely 1989), from which it can be inferred that the activation of these genes implies physiological conditions of the blood, a scenario that is expected to occur during a disease process (Perez et al. 2009).

6. BACTERIAL RESPONSES TO OSMOTIC CHALLENGES

The analysis of the bacterial response to osmotic stress has been studied in a variety of microorganisms (bacteria, archaea, and single-celled eukaryotes) Among the bacterial species that have been studied the most in response to this stress are *E. coli*, an inhabitant of the intestinal tract of most mammals, *Bacillus subtilis* and *Corynebacterium glutamicum*, very frequent inhabitants of the soil, and *Halomonas* has been the most studied species in ecosystems with high salt concentrations (Wood, 2011). Several authors posit that bacteria have integrated aquaporin membranes, or proteins related to the regulation of water balance (Tamma et al. 2018).

Additionally, bacteria are surrounded by a rigid and elastic structure, the cell wall (made up of different polymers such as murein or peptidoglycan), which give a particular morphology to each bacterium (Romaniuk & Cegelski 2015). Finally, it is necessary to mention the periplasmic space, which is located between the wall and the membrane, and which is substantially different between gram-positive and gram-negative bacteria. All these structures must maintain an integrity based on hydration that is also directly a function of the content of solutes and, therefore, the osmotic pressure derived from these conditions (Altendorf et al. 2009). The enzymatic systems that regulate the internal concentration of solutes are varied, however, those described for *E. coli* are of special relevance, since being present in the intestinal tract, the modifications in the concentrations of the different solutes are highly variable, as they are dependent on the diet. This resulted in adaptive effects for the bacterium, developing multiple and redundant mechanisms to maintain homeostatic balance in cellular physiology, under such fluctuating environmental conditions. Several examples of the mechanisms involved in this function are shown in Figure 2.

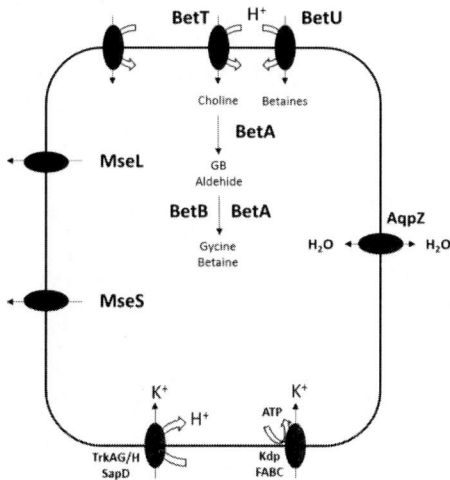

Figure 2. Principals mechanism involve on osmotic regulation.

7. Aquoporins and Osmostress: Water Balance and Adequacy

The discovery of aquoporins received the Nobel Prize (Calamita et al. 1995). Aquoporins are proteins that allow cells to regulate the water balance and, therefore, it is one of the most relevant mechanisms in cellular physiology that allows regulating the molar concentrations of different elements in the cytoplasm. These regulatory processes are based on the accumulation of organic osmolytes and fluctuate when environmental osmolality increases (Bremer and Krämer 2000; Csonka 1989; Kempf and Bremer 1998; Wood 2011). Expulsion of these osmolytes and other metabolites occurs through a mechanism based on mechanosensitive channels when the environmental osmolality is reduced (Booth 2014; Booth 2012; Cox et al. 2018).

8. General Elements That Characterize Osmoregulating Systems

1) The cell must be able to detect osmostress qualitatively and quantitatively. These mechanisms allow the cell to modulate the response based on environmental conditions, such as the accumulation of certain solutes.
2) Cells are able to change (activate/cancel) stress response systems according to environmental conditions. In this sense, regulatory phenomena are related to the effectiveness of the response in terms of the activation speed, as well as the period of time for which that system is operating.
3) The systems for the osmolytes detection are regulated based on the carbon/energy ratio. Therefore, homeostatic regulation is focused on optimizing resources in terms of energy efficiency (Bremer and Krämer 2000; Kempf & Bremer 1998; Wood et al. 2001).

CONCLUSION

The works reviewed in this chapter allow us to infer that the plasma membrane is a structure with different functions (active and passive) that plays a central role in cellular physiology and is particularly relevant for the maintenance of homeostatic regulation. Homeostatic regulation in the membrane must be understood fundamentally within two major cellular physiology processes: the perception of environmental conditions, which are limited specifically to the entire variety of cellular receptors, which, through transduction chains, allow the cell to generate the most effective response, and the transport of substances between the exterior and the interior of the cell, the most general physiological effect. The latter can be understood in terms of maintaining the control of molar concentrations in a range that enables optimal functioning of all the molecular elements in the cell. The membrane, based on its general distribution in all cells of the biosphere, must have developed at the origin of life, because homeostatic control is the core for the development of biological systems. The membrane is the most relevant structure that combines non-specific adaptation mechanisms responsible for bacterial adaptation and, later on, for eukaryotes adaptation to ecosystem conditions.

REFERENCES

Aendekerk, S., Ghysels, B., Cornelis, P. & Baysse, C. (2002). Characterization of a new efflux pump, MexGHI-OpmD, from *Pseudomonas aeruginosa* that confers resistance to vanadium. *Microbiology* 148: 2371–2381. doi: 10.1099/00221287-148-8-2371.

Altendorf, K., I Booth, R. Gralla, J. D., Greie, J. C. Rosenthal, A. Z. & Wood. J. M. (2009). Osmotic stress. *Eco Sal Plus*. doi: 10.1128/ecosalplus.5.4.5.

Angelova, M. I., Bitbol, A. F., Seigneuret, M., Staneva, G., Kodama, A., Sakuma, Y., Kawakatsu, T., Imai M., & Puff, N. (2018). pH sensing by lipids in membranes: The fundamentals of pH-driven migration, polarization and deformations of lipid bilayer assemblies, *Biochimica et Biophysica Acta* (BBA) - *Biomembranes,* 1860, 10, 2042-2063. doi: 10.1016/j.bbamem.2018.02.026.

Ballweg, S. & Ernst, R. (2017). Control of membrane fluidity: The OLE pathway in focus. *Biol Chem,* 398:215-228. doi: 10.1515/hsz-2016-0277

Blanc-Potard, A. B. & Groisman, E. A. (1997). The *Salmonella* selC locus contains a pathogenicity island mediating intramacrophage survival. *EMBO J.*; 16:5376–85. https://dx.doi.org/10.1093%2Femboj%2F16.17.5376.

Blanco, P., Hernando-Amado, S., Reales-Calderon, J. A., Corona, F., Lira, F., Alcalde-Rico, M., Bernardini, A., Sanchez, M. B., & Martinez, J. L. (2016). Bacterial Multidrug Efflux Pumps: Much More Than Antibiotic Resistance Determinants. *Microorganisms,* 4(1), 14. doi: 10.3390/microorganisms4010014.

Brandt, K. K., Petersen, A., Holm, P. E. & Nybroe, O. (2006). Decreased abundance and diversity of culturable Pseudomonas spp. populations with increasing copper exposure in the sugar beet rhizosphere. *FEMS Microb Ecol.*; 56: 281-291. https://doi.org/10.1111/j.1574-6941.2006.00081.x.

Bremer, E. & Krämer, R. (2000). Coping with osmotic challenges: osmoregulation through accumulation and release of compatible solutes. In *Bacterial Stress Responses,* ed. G. Storz, R. Hengge-Aronis, 79–97. Washington, DC: ASM doi: 10.1016/S1095-6433(00)80031-8.

Booth, I. R. (2014). Bacterial mechanosensitive channels: progress towards an understanding of their roles in cell physiology. *Curr. Opin. Microbiol.* 18:16–22. doi: 10.1016/j.mib.2014.01.005.

Booth, I. R. & Blount, P. (2012). The MscS and MscL families of mechanosensitive channels act as microbial emergency release valves. *J. Bacteriol.* 194:4802–9. doi: 10.1128/JB.00576-12.

Calamita, G., Bishai, W. R., Preston, G. M., Guggino, W. B. & Agre, P. (1995). Molecular cloning and characterization of AqpZ, a water channel from *Escherichia coli. J. Biol. Chem.* 270:29063–66. doi: 10.1074/jbc.270.49.2906.

Cho, U. S., Bader, M. W., Amaya, M. F., Daley, M. E., Klevit, R. E., Miller, S. I. & Xu, W. (2006). Metal bridges between the PhoQ sensor domain and the membrane regulate transmembrane signaling. *J Mol Biol*.; 356:1193–206. https://doi.org/10.1016/j.jmb.2005.12.032.

Chua, N. K., Howe, V., Jatana, N., Thukral, L. & Brown, A. J. (2017). A conserved degron containing an amphipathic helix regulates the cholesterol-mediated turnover of human squalene monooxygenase, a rate-limiting enzyme in cholesterol synthesis. *J Biol Chem*, 292:19959-19973. doi: 10.1074/jbc.M117.794230.

Cornell, R. B. (2016). Membrane lipid compositional sensing by the inducible amphipathic helix of CCT. *Biochim Biophys Acta – Mol Cell Biol Lipids*, 1861:847-861. doi: 10.1016/j.bbalip.2015.12.022.

Covino, R., Ballweg, S., Stordeur, C., Michaelis, J. B., Puth, K., Wernig, F., Bahrami, A., Ernst, A. M., Hummer, G. & Ernst, R. (2016). A eukaryotic sensor for membrane lipid saturation. *Mol Cell*, 63:49-59. doi: 10.1016/j.molcel.2016.05.015.

Cox, C. D., Bavi, N. & Martinac, B. (2018). Bacterial mechanosensors. *Annu. Rev. Physiol*. 80:71–93. doi: 10.1146/annurev-physiol-021317-121351.

Cromie, M. J., Shi, Y., Latifi, T. & Groisman, E. A. (2006). An RNA sensor for intracellular Mg2+ *Cell*.; 125:71–84. https://doi.org/10.1016/j.cell.2006.01.043.

Cronan, J. E. Jr & Gelmann, E. P. (1975). Physical properties of membrane lipids: biological relevance and regulation. *Bacteriol. Rev*. 39, 232–256. PMCID: PMC413917.

Csonka, L. N. (1989). Physiological and genetic responses of bacteria to osmotic stress. *Microbiol. Rev*. 53:121–47. PMC372720.

Cybulski, L. E., Martín, M., Mansilla, M. C., Fernández, A. & De Mendoza, D. (2010) Membrane thickness cue for cold sensing in a bacterium. *Curr Biol*, 20:1539-1544. doi: 10.1016/j.cub.2010.06.074.

Dann, C. E., Wakeman, C. A., Sieling, C. L., Baker, S. C. & Irnov, I. Winkler W. C. (2007). Structure and mechanism of a metal-sensing regulatory RNA. *Cell*; 130:878–92. https://doi.org/10.1016/j.cell.2007.06.051.

Davies, J. & Davies, D. (2010). Origins and evolution of antibiotic resistance. *Microbiology and molecular biology reviews: MMBR*, *74*(3), 417–433. doi: 10.1128/MMBR.00016-10.

Ernst, R., Ejsing, C. S. & Antonny, B. (2016). Homeoviscous adaptation and the regulation of membrane lipids. *J Mol Biol*, 428:4776- 4791. 8.

Falke, J. J. (2007). Membrane Recruitment as a Cancer Mechanism: A Case Study of Akt PH Domain. *Cell science*, *4*(2), 25–30. PMCID: PMC2601639.

Freeman, J. L., Persans, M. W., Nieman, K., & Salt, D. E. (2005). Nickel and cobalt resistance engineered in Escherichia coli by overexpression of serine acetyltransferase from the nickel hyperaccumulator plant *Thlaspi goesingense. Applied and environmental microbiology*, 71(12), 8627–8633. https://dx.doi.org/10.1128%2FAEM.71.12.8627-8633. 2005.

Garcia Vescovi, E., Soncini, F. C. & Groisman, E. A. (1996). Mg2+ as an extracellular signal: environmental regulation of *Salmonella* virulence. *Cell.;* 84:165–74. https://doi.org/10.1016/S0092-8674(00)81003-X.

Gatenby, R. A. (2019). The Role of Cell Membrane Information Reception, Processing, and Communication in the Structure and Function of Multicellular Tissue. *International journal of molecular sciences*, *20*(15), 3609. https://doi.org/10.3390/ijms20153609.

Götz, F. (2002). *Staphylococcus* and biofilms. *Molecular Microbiology*, 43: 1367-1378. https://doi.org/10.1007/978-3-540-75418-3_10.

Groisman, E. A., Hollands, K., Kriner, M. A., Lee, E. J., Park, S. Y., & Pontes, M. H. (2013). Bacterial Mg2+ homeostasis, transport, and virulence. *Annual review of genetics*, 47, 625–646. doi: 10.1146/annurev-genet-051313-051025.

Guha, C. & Mookerjee, A. (1981) RNA synthesis and degradation during preferential inhibition of protein synthesis bycobalt chloridein Escherichia coli K-12. *Mol Biol Rep* 7: 217-220. https://doi.org/10.1007/BF00805755.

Hazel, J. R. (1995). Thermal adaptation in biological membranes is homeoviscous adaptation the explanation. *Annu Rev Physiol*, 57:19-42. https://doi.org/10.1146/annurev.ph.57.030195.000315.

Henkin, T. M. (2008). Riboswitch RNAs: using RNA to sense cellular metabolism. *Genes Dev.*; 22:3383–90. https://dx.doi.org/10.1101%2Fgad.1747308.

Hmiel, S. P., Snavely, M. D. & Miller, C. G., Maguire, M. E. (1986). Magnesium transport in *Salmonella typhimurium*: characterization of magnesium influx and cloning of a transport gene. *J Bacteriol.*; 168:1444–50.

Hmiel, S. P., Snavely, M. D., Florer, J. B., Maguire, M. E. & Miller, C. G. (1989). Magnesium transport in *Salmonella typhimurium*: genetic characterization and cloning of three magnesium transport loci. *J Bacteriol.*; 171:4742–51. https://dx.doi.org/10.1128%2Fjb.171.9.4742-4751.1989.

Kempf, B. & Bremer, E. (1998). Uptake and synthesis of compatible solutes as microbial stress responses to high osmolality environments. *Arch. Microbiol.* 170:319–30. https://doi.org/10.1007/s002030050649.

Krul, J. (1977). Some factors affecting floc formation by Zoogloearamigera, strainI-16-M. *Water Res.*; 11: 51-56. https://doi.org/10.1016/0043-1354(77)90181-6.

Lee, D. G., Urbach, J. M., Wu, G., Liberati, N. T., Feinbaum, R. L., Miyata, S., Diggins, L. T., He, J., Saucier, M., Déziel, E., Friedman, L., Li, L., Grills, G., Montgomery, K., Kucherlapati, R., Rahme, L. G. & Ausubel, F. M. (2006). Genomic analysis reveals that *Pseudomonas aeruginosa* virulence is combinatorial. *Genome Biol.*; 7(10):R90. doi: 10.1186/gb-2006-7-10-r90.

Lee, E. H., Rouquette-Loughlin, C., Folster, J. P. & Shafer, W. M. (2003) FarR regulates the farAB-encoded efflux pump of *Neisseria gonorrhoeae* via an MtrR regulatory mechanism. *J Bacteriol* 185: 7145–7152. doi: 10.1128/JB.185.24.7145-7152.2003.

Lewinson, O. & Bibi, E. (2001) Evidence for simultaneous binding of dissimilar substrates by the *Escherichia coli* multidrug transporter MdfA. *Biochemistry* 40: 12612–12618. https://doi.org/10.1021/bi011040y.

Li, X. Z., & Nikaido, H. (2009). Efflux-mediated drug resistance in bacteria: an update. *Drugs*, 69(12), 1555–1623. doi: 10.2165/11317030-000000000-00000.

Martinez, J. L., M. B., Sánchez, Martínez-Solano,L., Hernandez, A., Garmendia, L., Fajardo, A., & Alvarez-Ortega,C. (2009)Functional role of bacterial multidrug efflux pumps in microbial natural ecosystems, *FEMS Microbiology Reviews*, 33, 2, Pages 430–449. doi: 10.1111/j.1574-6976.2008.00157.x.

Miller M. B. & Bassler B. L. Quorum Sensing in Bacteria *Annual Review of Microbiology* 2001 55:1, 165-199 doi: 10.1146/annurev.micro.55.1.165.

Neyfakh, A. A. (1997). Natural functions of bacterial multidrug transporters. *Trends Microbiol* 5: 309–313. https://doi.org/10.1016/S0966-842X(97)01064-0.

Nies, D. H. (2003). Efflux-mediated heavy metal resistance in prokaryotes. *FEMS Microbiology Reviews*, 27: 313-339. doi: 10.1016/S0168-6445(03)00048-2.

O'Hara, L., Han, G. S., Sew, P. C., Grimsey, N., Carman, G. M. & Siniossoglou, S. (2006). Control of phospholipid synthesis by phosphorylation of the yeast lipin Pah1p/Smp2p Mg2 + dependent phosphatidate phosphatase. *J Biol Chem*, 281:34537-34548. https://dx.doi.org/10.1074%2Fjbc.M606654200.

Perez, J. C., Shin, D., Zwir, I., Latifi, T. Hadley, T. J. & Groisman, E. A. (2009). Evolution of a bacterial regulon controlling virulence and Mg2+ homeostasis. *PLoS Genet*.; 5:e1000428. https://doi.org/10.1371/journal.pgen.1000428.

Perron, K., Caille, O., Rossier, C., Van Delden, C., Dumas, J. L. & Kohler, T. (2004). CzcR–CzcS, a two-component system involved in heavy metal and carbapenem resistance in *Pseudomonas aeruginosa*. *J Biol Chem* 279: 8761–8768. doi: 10.1074/jbc.M312080200.

Robinson, J. B., & Tuovinen, O. H. (1984). Mechanisms of microbial resistance and detoxification of mercury and organomercury compounds: physiological, biochemical, and genetic analyses. *Microbiological reviews*, 48(2), 95–124. PMCID: PMC373215.

Romaniuk, J. A. & Cegelski, L. (2015). Bacterial cell wall composition and the influence of antibiotics by cell-wall and whole-cell NMR. *Philosophical transactions of the Royal Society of London. Series B, Biological sciences*, *370*(1679), 20150024.

Roth, A. & Breaker, R. R. (2009). The structural and functional diversity of metabolite-binding riboswitches. *Annu Rev Biochem.*; 78:305–34. doi: 10.1146/annurev.biochem.78.070507.135656.

Saier Jr, M. H., Paulsen, I. T., Sliwinski, M. K., Pao, S. S., Skurray, R. A. & Nikaido, H. (1998) Evolutionary origins of multidrug and drug-specific efflux pumps in bacteria. *FASEB J* 12: 265–274. https://doi.org/10.1096/fasebj.12.3.265.

Saita, E., Albanesi, D. & De Mendoza, D. (2016). Sensing membrane thickness: lessons learned from cold stress. *Biochim Biophys Acta – Mol Cell Biol Lipids*, 1861:837-846. doi: 10.1016/j.bbalip.2016.01.003.

Sekiya, H., Mima, T., Morita, Y., Kuroda, T., Mizushima, T. & Tsuchiya, T. (2003). Functional cloning and characterization of a multidrug efflux pump, mexHI–opmD, from a *Pseudomonas aeruginosa* mutant. *Antimicrob Agents Ch* 47: 2990–2992. https://dx.doi.org/10.1128%2FAAC.47.9.2990-2992.2003.

Sherif, M, Eman E. S., Eman A. F., Kelany, A, & Rizk, R. (2011). Effects of acute low doses of Gamma-radiation on erythrocytes membrane. *Radiation and environmental biophysics*. 50. 189-98. 10.1007/s00411-010-0333-x.

Silver, S. & Phung, L. T. (1996). Bacterial heavy metal resistance: new surprises. *Annu Rev Microbiol* 50: 753–789. doi: 10.1146/annurev.micro.50.1.753.

Silver, S. & Phung, L. T. (2005). A bacterial view of the periodic table: genes and proteins for toxic inorganic ions. *J Ind Microbiol Biot* 32: 587–605. https://doi.org/10.1007/s10295-005-0019-6.

Singh, R., Gautam, N., Mishra, A., & Gupta, R. (2011). Heavy metals and living systems: An overview. *Indian journal of pharmacology*, *43*(3), 246–253. doi: 10.4103/0253-7613.81505.

Sinensky. M. (1974) Homeoviscous adaptation a homeostatic process that regulates the viscosity of membrane lipids in *Escherichia coli. Proc Natl Acad Sci* USA, 71:522-525. https://doi.org/10.1073/pnas.71.2.522.

Sinensky, M. (1980) Adaptive alteration in phospholipid composition of plasma membranes from a somatic cell mutant defective in the regulation of cholesterol biosynthesis. *J Cell Biol*, 85:166-169.

Smith, R. L., Thompson, L. J. & Maguire, M. E. (1995). Cloning and characterization of MgtE, a putative new class of Mg2+ transporter from *Bacillus firmus* OF4. *J Bacteriol*.; 177:1233–38. https://dx.doi.org/10.1128%2Fjb.177.5.1233-1238.1995.

Snavely, M. D., Florer, J. B., Miller, C. G. & Maguire, M. E. (1989). Magnesium transport in *Salmonella typhimurium*: 28Mg2+ transport by the CorA, MgtA, and MgtB systems. *J Bacteriol*.; 171:4761–66. https://dx.doi.org/10.1128%2Fjb.171.9.4761-4766.1989.

Sunshine, H. & Iruela-Arispe, M. L. (2017). Membrane lipids and cell signaling. *Current opinion in lipidology*, 28(5), 408–413. doi: 10.1097/MOL.0000000000000443.

Tamma, G., Valenti, G., Grossini, E., Donnini, S., Marino, A., Marinelli, R. A., & Calamita, G. (2018). Aquaporin Membrane Channels in Oxidative Stress, Cell Signaling, and Aging: *Recent Advances and Research Trends. Oxidative medicine and cellular longevity*, 1501847. https://doi.org/10.1155/2018/1501847.

Teitzel, G. M., Geddie, A., De Long, S. K., Kirisits, M. J., Whiteley, M. & Parsek, M. R. (2006). Survival and growth in the presence of elevated copper: transcriptional profiling of copper-stressed *Pseudomonas aeruginosa. J Bacteriol* 188: 7242–7256. doi: 10.1128/JB.00837-06.

Tulodziecka, K., Diaz-Rohrer, B. B., Farley, M. M., Chan, R. B., Di Paolo, G., Levental, K. R., Waxham, M. N. & Levental, I. (2016). Remodeling of the postsynaptic plasma membrane during neural development. *Mol Biol Cell*, 27:3480-3489. doi: 10.1091/mbc.E16-06-0420.

Wood, J. M. (2015) Bacterial responses to osmotic challenges. *J Gen Physiol* 1 May; 145 (5): 381–388. doi: 10.1085/jgp.201411296.

Wood, J. M. (2011). Chapter 9: Osmotic Stress. In *Bacterial Stress Responses*. G. Storz, and R. Hengge, editors. ASM Press, Washington, D.C. 133–156. https://dx.doi.org/10.1085%2Fjgp.201411296.

Wood, J. M. (2011). Bacterial osmoregulation: a paradigm for the study of cellular homeostasis. *Annu. Rev.Microbiol.* 65:215–38.

Wood, J. M., Bremer, E., Csonka, L. N., Krämer, R., Poolman, B, van der Heidee, T, Smithf L. T. (2001). Osmosensing and osmoregulatory compatible solute accumulation by bacteria. *Comp. Biochem. Physiol. A Mol. Integr. Physiol.* 130:437–60. https://doi.org/10.1016/S1095-6433(01)00442-1.

Zhang, Y. & Rock, C. (2008). Membrane lipid homeostasis in bacteria. *Nat Rev Microbiol* **6,** 222–233. https://doi.org/10.1038/nrmicro1839.

Zhou, H. & Jin, W. (2018). Membranes with Intrinsic Micro-Porosity: Structure, Solubility, and Applications. *Membranes*, *9*(1), 3. https://doi.org/10.3390/membranes9010003.

In: Molecular Basis of Specific Mechanism ... ISBN: 978-1-53618-751-9
Editors: Marcos López-Pérez et al. © 2020 Nova Science Publishers, Inc.

Chapter 5

LACCASES, A PROTEIN SYSTEM FOR ADAPTATION THROUGH THE USE OF RECALCITRANT RESOURCES: MOLECULAR BASIS AND COMPUTATIONAL MODELING APPROACHES TO USES IN THE BIOINDUSTRIES

*E. Villegas[1], A. Trejo-Martínez[2]
and L. D. Herrera-Zúñiga[1,2],**

[1]Área de Estudios de Posgrado e Investigación,
Tecnológico de Estudios Superiores del Oriente del Estado de México
[2]Laboratorio de Estructura y Función de Proteínas,
Centro de Investigación en Biotecnología,
Universidad Autónoma del Estado de Morelos

ABSTRACT

Laccases are the oldest and most studied enzymes since they were discovered in *Rhus vernicifera* tree in 1920. These enzymes have

---

* Corresponding Author's Email: leonardo.herrera@tesoem.edu.mx.

regaining popularity due to its use in environmental and biotechnological systems. In this work, several biochemical properties of laccases are reviewed from computational points of view. In this review, a structural laccase overlapping was performed over PDB crystallographic proteins to demonstrate that the trinuclear center remains intact among them. Domains 1 and 3 are strongly conserved among all Small and three domains laccases. Moreover, a more precise structure overlapping on the catalytic site suggested that two domain laccases lost domain number 2, which is still present in three-domain laccases. Domain number 2 has evolutionarily been lost to form the activated protein. In conclusion, trinuclear cluster present in two-domain laccases to conform the functional homotrimer as it has been confirmed in several phylogenetic studies. This chapter present, the state-of-the-art review of modeling methods such as molecular modeling and dynamics simulations, quantum and hybrid approaches that may help to understand laccase structures, interactions, biocatalysis and protein engineering to be used for biotechnological applications.

Keywords: docking, *in-silico*, molecular dynamics, QM-MM

1. INTRODUCTION

The p–benzenediol: oxygen oxidoreductase EC 1.10.3.2, commonly known as laccase enzyme is a multicopper oxidase containing copper in its active site, they catalyze the oxidation of a broad range of compounds as phenol and amines. Laccase can reduce molecular oxygen to water without producing harmful products (Janusz et al., 2020). This enzyme is a monomeric protein, highly glycosylated with a molecular weight around 50-140 kDa (Arregui et al., 2019). Laccase is commonly present in fungi, plants, insects and bacteria, it is involved in a wide range of biological processes from pigment formation in fungi like *Pycnoporus sanguineus* (Acosta-Urdapilleta et al., 2010) to sexual development in *Schizophyllum commune* (Madhavan et al., 2014). Industrial operations are conducted under extreme conditions, such as, elevated temperatures, extreme pH, high salt concentrations or in the presence of organic solvents, where fungal laccases usually do not operate at such as in extreme conditions (Arregui et al., 2019). Even though, fungal laccases are versatile enzymes which activities can be found in a wide range of temperature and pH, i.e., *Tricholoma matsutake* and *Aquifex aeolicus* lacasses presented higher enzymatic activity at 20 and 90°C, respectively (Devi et al., 2016; Hildén et al., 2009; Toledo-Núñez et al., 2012), while growing *Trametes hirsuta* tolerated temperature from 4 to 48°C ± 2 and pH 3-13 (Dhakar et al., 2013); however, fungal laccases can be inactivated at neutral and alkaline pH, even thought, they are structurally stable above 7.0. In contrast, bacterial laccases are more stable at the same pH conditions. So, an interesting aspects of fungal laccase catalysis is its pH dependence. Most fungal laccases display an optimum pH activity from 3.0 to 5.5 against phenolic substrates and are basically inactive as pH is approaching to neutral and alkaline, although they are structurally stable above 7.0. Engineering fungal laccase Lcc9 from *Coprinopsis cinereal* a new laccase PIE5 was constructed to improve its enzymatic activity at pH 8.5 for indigo dye decolorization. PIE5 has three amino acid modifications (E116 K, N229D, I393 T) when expressed in *Pichia pastoris*. Moreover, PIE5 site-directed mutagenesis showed that N229D replacement contributed to a better tolerance of pH, causing a 1.5-unit increase in pH. Thus, computational engineering improves the performance of alkaline laccases and some applications could be directed to: a) The production of bioethanol by steam-exploded wheat straw as a substrate with laccase at pH 8 decreases the lignin content of the solid fraction and increases the production of glucose and xylose after saccharification, and b) In the hair coloring industries laccases at pH 9 are of common applications. Neutral lacasses with high activity at blood pH 7.4 have potential to be used in implantable self-contained wireless 3D nano-biodevices that work in different physiological fluid. Moreover, most laccases known to date have optimal activity at acidic pH, which can prohibit their use in systems operating at alkaline conditions. In the food industry, is used in beverages stabilization, fruit juice processing, baking, improving of food sensory parameters, sugar beet pectin gelation (Osma et al., 2010); in textile industry is to bleach denim, cotton, dye cotton and wool, synthesis of novel phenazine and phenoxazinone dyes, eliminate the odor in cloth washing and in the shrinkage of wool (Mojsov et al., 2014); in paper and pulp industry, is used to transform lignin to lower aromatic compounds, to remove color and stabilizing effluents from paper mills, old newsprint and reused and biografting of pulp fibers (Singh et al., 2019, Singh et al., 2019); in pharmaceutical chemistry, for anticancer, antibiotics, antifungal drugs development, prostaglandin production, melanin synthesis, sedatives and in detoxifying the toxic compounds (Chaurasia, et al., 2016); enzymatic bioremediation, laccase can catalyze the degradation of xenobiotics, decolorization of dyes effluent treatment and be used in biofuels production, biosensor, DNA labeling, immunochemical assay, bioorganic compound synthesis cosmetics; paints, coatings, agrochemicals, and furniture (Arregui et al., 2019; Benson et al., 1990; Desai et al., 2011; Kudanga et al., 2014; Mate et al, 2017; Moreno, Ibarra, Eugenio et al., 2020; Viswanath et al., 2014). In recent years, due to its high industrial potential, laccase has attracted the attention of scientists, and a higher number of research papers has been published, in the last decades The numbers have steadily risen from a few in the 1990 to more than 300 to the date. (Figure 1). Most of this research has been focused on fungal laccases, although they have also been studied on plants, insects, bacteria and even archaea.

There were attempts to manufacture large amounts of laccases using recombinant organisms or screening for natural hypersecretory strains. The two most popular approaches to these issues are the discovery of novel enzymes with desired properties, and genetic engineering for the modification of the properties of known counterparts. The adaptation of the catalytic laccases properties has been also carried out by logical design and driven evolution, mostly led by computational simulations. there are several articles where computational methods have been used to analyze sequences data, identified genes and proteins, to determine structural characteristics, phylogenetic analysis, find key residues in interaction with specific amino acids by docking, that allows to suggest amino acids point changes for targeted mutagenesis experiments, to improve pH activity in alkaline conditions by molecular dynamics or quantum mechanics to calculate the redox potential. This chapter aim is to open-up new possibilities to the study laccase enzyme properties by chemistry and biology computational methodologies. The next section summarizes the state of the art of computational experiments that has been published to study laccase enzyme, starting from structural analysis, in-silico and light computing approaches, modeling, docking, mechanics and molecular dynamics, to arrive at the methodologies of quantum mechanics

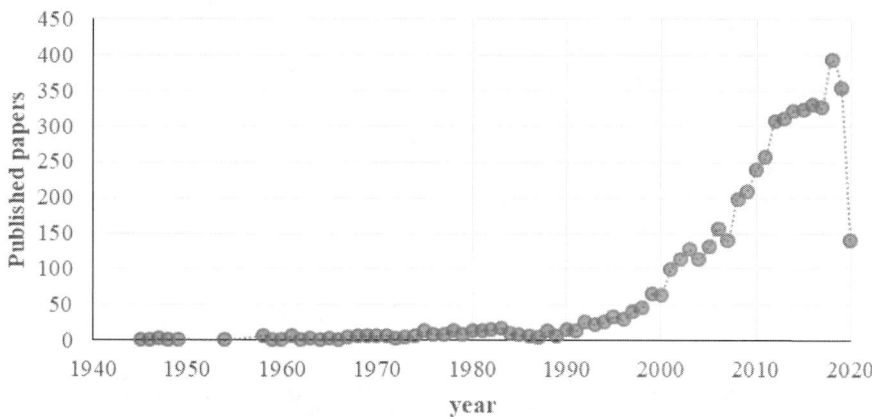

Figure 1. Laccase annual number publications. Data registered by the National Center for Biotechnology Information (NCBI) until March of 2020 (Benson et al., 1990).

1.1. Structural Analysis of Cooper Evolution in Laccases

Laccases have four copper atoms forming their catalytic site named Tetranuclear Cluster, TNC, (Figure 2A) in ochre circles. Common laccases are formed by three homologous superstructures, domains or motif, each one of them constituted in seven antiparallel β-strands, twisted and coupled in two β-sheets, arranged into a closed barrel structure, named Greek-key barrel motif or cupredoxin domains found in the green alga plastocyanin. Their catalytic site TNC is built by one mononuclear copper site (T1) wrapped in domain 3 and the trinuclear cluster formed at the interface between domain 1 and 3, domain 2 has been removed for clarity, (Figure 2B).

88   E. Villegas, A. Trejo-Martínez and L. D. Herrera-Zúñiga

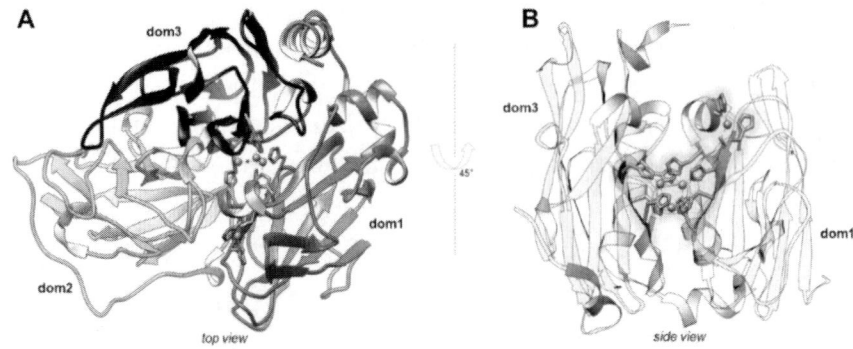

Figure 2. Structural insight into laccases from *Steccherinum murashkinskyi*, PDB-ID: 6UI6. (A) Laccase three domains, 1 in orange, 2 in cyan and 3 in blue with four coppers (Ochre circles) and (B) TNC or catalytic site, architecture and structural features highlighted of the copper-binding without domain two.

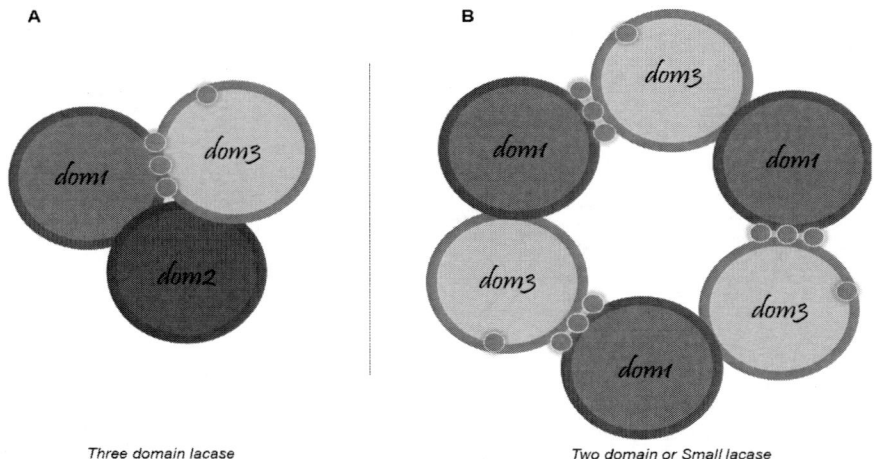

Figure 3. Schematic structural localization diagram of TNC between two and three domain laccases. A) Three domain laccase monomer. B) Two domain laccase homotrimer. In blue domain1, in red domain 2 and in green domain 3, cuppers in ochre circles.

By 1998 Drucos et al., elucidated the first laccase structure crystal from the *Coprinus cinereus* fungus, with three domains of cupredoxin (Ducros et al., 1998). Laccases are composed by three-domain or six-domain (small laccases). In three domains laccase, T1 site is in domain 3 and the three nuclear cluster is located at the interface between domains 1 and 3 Figure 3A. T1 site in two domain laccases or small laccase is inside 2nd, 4th and 6th domains, but for the formation of the trinuclear cluster they need to oligomerize as homotrimers generating three nuclear cluster in the interface between $1^{st}/6^{th}$, $2^{nd}/3^{rd}$ and $4^{th}/5^{th}$ domains among monomers Figure 3B. Understanding that two domain laccases or small laccase or six domains laccases have not a small structure, neither an easy name to help to remember its structure. Several hypotheses on the evolution of laccases domains and related to copper have been proposed; all of them consider that the cupredoxin domain, with one copper atom in its structure, developed in different forms of three-domain laccases. Domains one and three house the copper sites, and the second domain also helps to form a cleft that connects the substrates.

In 2004 Canters et al., found a new laccase structure, "The small laccase" with two domains from *Streptomyces coelicolor*, different from three domain since Its genome encode a small, four-copper oxidase, which lacks the second domain. This protein is representative of a new enzyme family-the two-domain laccases (Machczynski et al., 2004). The overlap between plastocyanin and three domain laccase structures suggest a high probability that laccase evolved from plastocyanin which is the minimal structural unit, one domain with one atom copper, commonly named copper type - 1. An overlapping of the protein structure plastocyanin and laccase three domains is shown in Figure 4. In Green, the plastocyan in from a green alga *Enteromorpha prolifera* (PDB-ID: 7PCY), and in red color the *Bacillus subtilis* laccase enzyme (PDB-ID: 5ZJI). In red, orange and orange light laccase domains, the RMSD of alpha-carbons in the superposition of plastocyanin and dom3-laccase are close to 1 Å, Figure 4, for structural details of this or other proteins, visit PDBsum, available at www.ebi.ac.uk/pdbsum/, (Laskowski et al., 2018).

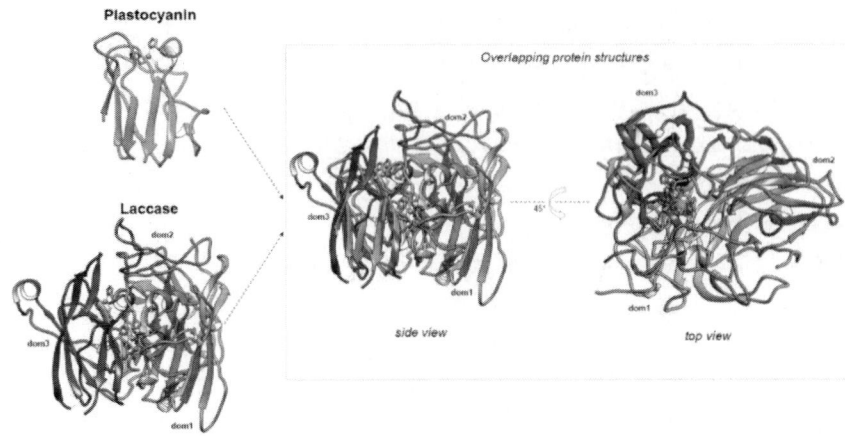

Figure 4. Overlapping protein structure plastocyanin and laccase three domains.

Just a limited number of publications are dedicated specifically to the evolution of laccases Rydén et al.,1993, Valderrama et al., 2003, Komori et al., 2010, Vasin et al., 2013, Janusz et al., 2020, who have suggested a common ancestor for laccase, they all proposed that domain cupredoxin with one copper atom in its structure, formed in different structure patterns of the multicopper oxidase enzyme family, including dicyanin, ascorbate oxidase, nitrite reductase, ceruloplasmine, small laccase and laccases of three domains (Janusz et al., 2020; Komori et al., 2009; Rydén & Hunt, 1993; Valderrama et al., 2003; Vasin et al., 2013). These hypotheses describe different pathways and intermediate species that have led to the development of the trinuclear cluster and the origin of the two different laccases. These two laccases maintain the copper-binding site and the interdomain association to preserve the tetra-nuclear cluster type of four copper atoms (Díaz-Godínez et al., 2018; Rubén-Díaz et al., 2018). The interesting assumption of this structural and catalytic site evolution of laccases hypothesis, is the structural conservation of TNC in two different laccases type: small laccases and three domain laccases which are showed in the schematic representation of homotrimers of small laccases. Figure 5A shows the side view (A1), the top view (A2), and the overlapping catalytic TNC site (1, 2 and 3) present in all PDB small laccases structures (Figure 5B, B1 Side view, B2 Top view).

Laccases, a Protein System for Adaptation ... 91

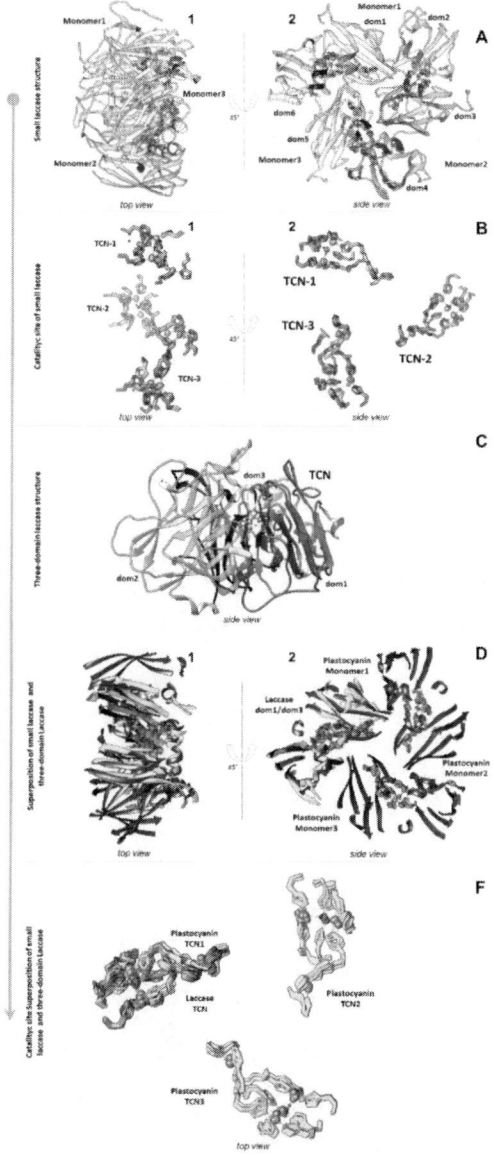

Figure 5. Structural conservation of de TNC in two different laccases types. A) Small laccases monomers structures, 1 in blue, 2 in pink, 3 in yellow. B) Catalytic site of small laccases, copper atoms are in ochre color, histidine: in blue nitrogen and gray carbon. C) Three laccase domains, 1 in orange, 2 in cyan and 3 in blue. D) Small laccase superposition structure and domains 1 and 3 of three laccase domain. F) Catalytic TNC superposition of laccase, in blue small laccases and yellow the three domain laccases.

A representative structure of three domain laccase is displayed in Figure 5C, and the overlapping structures of small laccases and of three domain laccase are exposed in Figure 5D (D1 side view, D2 top view). Also, the overlapping catalytic structure of small laccases and three domain laccase is shown in Figure 5E (E1 side view, E2 top view). Finally, the TNC structure overlapping each one of all structured laccases from different organisms depicted in the Protein Data Bank is presented in Figure 5F (F1 side view, F2 top view).

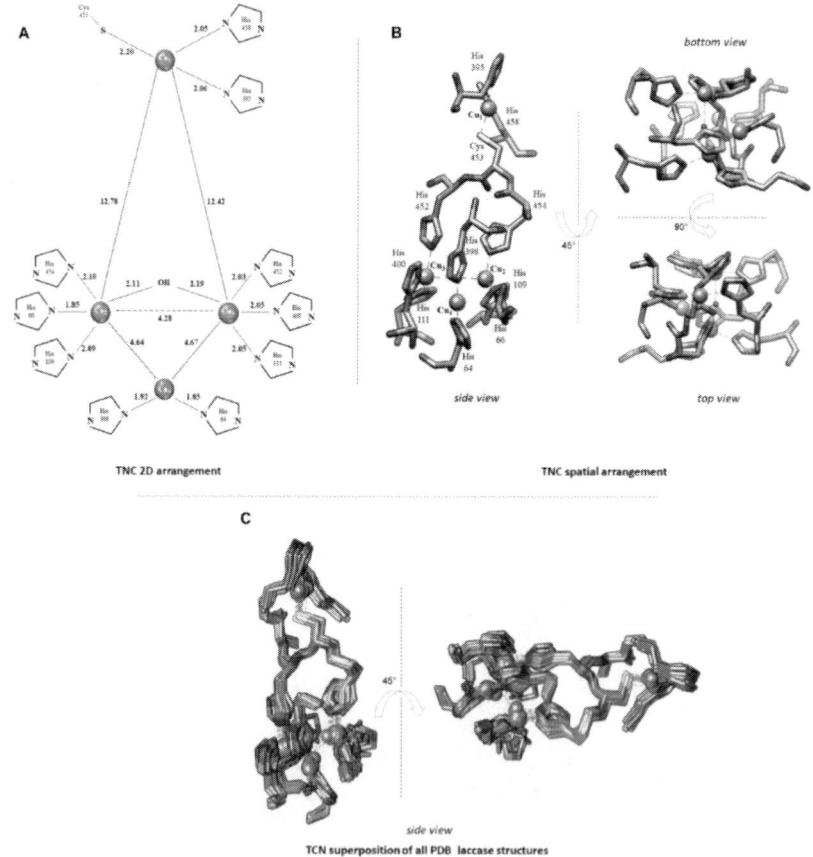

Figure 6. TNC spatial arrangement. A) Approximate distances of copper and its residue ligands, B) Structural copper coordination, and C) TNC structure overlapping of all structure laccases depicted in the Protein data Bank.

The three nuclear site is formed by one copper Type-2 (T2) and two coppers Type-3 (T3) coordinated only by histidine. T2 copper is rigorously coordinated by two histidine and copper T3 by six histidine. At the same time, two type T3 copper are coupling by a hydroxyl bridge. In both laccases, the approximate distance between copper Cu1 - Cu2 is 12.78 Å and 12.42 Å from Cu1 - Cu3 (Figure 6A). TCN spatial arrangement is displayed in Figure 6B. Superimposition analysis of all PDB laccases structures revealed a TCN root mean square deviation (RMSD) < 0.5 Å suggesting a TCN significant and high conservative scaffolding level among laccases, (Figure 6C).

Laccase structures depicted in the Protein Data Bank (PDB) preserving TNC are 41, of these 9 are two-domain and 32 are three domain laccases. The TI site shows the copper atom in the same trigonal orientation with two conserved histidine and one cysteine as equatorial ligands, and an axial ligand, the methionine in bacteria can be leucine or phenylalanine in fungus is of a variable nature.

We must point out the transcendence of the copper arrangement of the conservative evolution, but also its catalytic importance. The pioneer study of copper atoms and their scaffolding function was reported by Edward I. Solomon, in 1976. He understand the behavior of the type 3 copper in *Rhus vernicifera* laccase (https://web.stanford.edu/group/solomon/home.html), providing us with the knowledge of the copper-complex reactivity in a large number of copper-dependent proteins (Kim et al., 2019; Solomon et al., 1976). Solomon et al., have been one of the most assiduous users of computational techniques for the copper atoms studies, either using conventional-combined techniques such as quantum mechanical and molecular mechanics (QM/MM) or developing new techniques to combine calculations, like to experimental extended X-ray absorption fine structure (EXAFS) refinements with QM, EXAFS/QM (Ryde et al., 2007). In classic studies, coppers have been tamed for laccase, mostly with three force fields, Gromos, OPLS and Charmm, as it have been done by several authors such as, Christensen et al., 2013, Awasthi et al., 2015; Ferrario et al., 2015; Jones et al., 2015; Kepp et al., 2015; Herrera-Zuñiga et al., 2019; Maniak et al., 2020; Mehra et al., 2018; Monza et al., 2015. Moreover, copper reactivity of catalytic site has been study by Density Functional Theory, DFT, by de Salas et al., 2019 and Chaurasia et al., 2016, to understand copper ligands interaction, results indicated that the number of all atomic spin densities in the substrate indicates the total volume of unpaired (radical) electron due to its oxidation.

1.2. *In-Silico* and Light Computing Approaches

From the computational point of view of laccase, the most recent review was done by Piscitelli et al., in 2019, they showed that engineering methods through rational mutagenesis worked just fine to build highly efficient catalytic site in laccase (Stanzione et al., 2020). Rational, mechanistic design can significantly improve the performance of the bioremediation process for waste treatment and food safety. At the cellular level such improvement can be informed by theoretical observations, particularly of phenotype plasticity, cell signaling, and community assembly. At the molecular level, the authors suggested enzyme design using techniques such as Small Angle Neutron Scattering and Density Functional Theory, in a case-study of the interaction laccase with the food pollutant aflatoxin B1 (Zaccaria et al., 2020). Salas et al., in 2019, developed catalytic engineering by computational simulation to the crystal structure of 7D5 laccase (PDB 6H5Y), in *Saccharomyces cerevisiae* overproduced in *Aspergillus oryzae* due to heavier and heterogeneous glycosylation with a more oblate geometric structure. This enzyme presented superior catalytic constants towards all tested substrates, with no significant change in optimal pH or redox potential. They demonstrate a favorable binding and electron transfer from the substrate to the T1 copper due to the introduced V162A and E457D mutations (de Salas et al., 2019). Laccase application in biofuel and bioremediation allowed Kameshwar et al., in 2017, to compared and analyzed physico-chemical, structural and functional properties of fungal laccases from protein analysis to computationally predicted three-dimensional comparative models and to develop structurally and functionally stable laccase by a rapid and reliable protein pipeline to docking. Results showed that amino acids PRO, PHE, LEU, LYS and GLN in fungal laccases are strongly preserved around the active site and are used to interact with the ligands (Kameshwar et al., 2018). Engineering laccases, to improve the oxidation of small molecules for applications in multiple fields has been studied by Lucas et al., in 2017. They found that redesigning substrate binding at the T1 pocket, guided by *in silico* methodologies, for instance, Quantum Mechanics/Molecular Mechanics (QM/MM) is a consistent option. They evaluate the robustness of their computational approach to estimate activity, emphasizing the importance of the binding event in laccase reactivity. Strengths, weakness, potential for scoring large numbers of protein sequences and its significance in protein engineering were discussed and residues that could be susceptible to oxidation were identified by Spin Density values obtained from the calculations (Lucas et al., 2017). Avelar et al., 2018, in *Coriolopsis gallica* laccase identified residues that could be susceptible to oxidation by the Spin Density values. They simulated activated states of the catalytic region using QM/MM tools, after that, the results showed the electron distribution in both the basal and activated state (plus or minus one electron) of several conformations of three targets were selected (F357, F413 and F475) to replace by site-directed mutagenesis with less oxidizable residues such as leucine, alanine and isoleucine (Avelar et al., 2018). The surface computational engineering study by Robert et al., 2017 showed a total lysine reduction with unique lysine surface-accessible residue that make high efficiency olefin oxidation (Robert et al., 2017). Julio et al., 2017 demonstrated by structural computational biology laccase-insecticide interaction as an important factor for the resistance to pirimiphos-methyl and bifenthrin in *Tribolium castaneum* (Julio et al., 2017). Rational computational protein design to identify the most valuable structure for temperature resistance, demonstrate that the introduction of interface disulfide bonds in small laccases increased resistance to irreversible thermal denaturation (Garcia et al., 2016) and in an independent work Herrera-Zuñiga et al., 2019, in *Trametes versicolor* laccase, established that the introduction of interface salt bridges in three domain laccases surface can safeguard the three-dimensional structure of laccase up to high temperatures without effect on kinetic parameters (Herrera-Zúñiga et al., 2019). Structural modeling evolution and *in-silico* approaches were used by Herrera-Zuñiga et al., 2018, to design a new thermotolerant *Pleurotus ostreatus* laccase to demonstrate the best mutation in different location of the enzyme to make a thermotolerant mutation on the protein surface and promote the ionic interactions within it (Rubén Díaz et al., 2018). Molecular Modeling has been used to identify the role of laccase from the bacteria *Yersinia enterocolitica* in bioremediation of diclofenac and aspirin, to study the biotransformation and binding affinity of other emerging contaminants (Singh et al., 2016). Lu et al., in 2015, used Quantum Chemistry computational to demonstrated that laccase can couple products generated by reverse oxidation of halogenated phenols to humic acid (Lu et al., 2015). In three different works published in 2019, the stability of complexes laccase-dyes by docking and molecular dynamics was study to understand the low-redox activity of the white-rot fungus and bacterial laccases to clean effluents (Ahlawat et al., 2019; Ayla et al., 2019; Srinivasan et al., 2019). A study by molecular docking simulation between *Trametes versicolor* laccase and bisphenol A, demonstrated that the reaction laccase-ligand is spontaneous and disclose that BPA degradation flows across the carbon atoms to connect the two benzene rings of BPA molecule (Hongyan et al., 2019). They screened and used all laccases structures depicted in the Protein data Bank to determine its oxidation capability over endosulfan an insecticide molecule, using molecular docking and molecular dynamics simulation techniques, to found out that a bacterial *CotA* laccase, PDB-ID: 3ZDW from *Bacillus subtilis* has the best potential to degraded endosulfan (Singh et al., 2019). In the pharmacological industry laccase is used to accelerate studies of the coumarins derivates by molecular docking, suggesting that laccase is a reliable method to oxidative catalyze coupling to new active molecules i.e., Cinnabarinic acid an antibiotic molecule obtained by two molecules of 3 Hydroxy Anthranilic Acid, 3 HAA (Wang et al., 2019, and Eggert, 1995).

2. MOLECULAR MODELING STRATEGIES TO UNDERSTAND LACCASE STRUCTURE

The first molecular modeling of laccase was obyained by Cusanovich in 1986, to understand electron-transfer reactions between flavodoxin and flavins. In this work, the known three-dimensional structures for plastocyanin were used to demonstrate that the interaction site charge is similar between this two proteins (Tollin et al., 1986). Nowadays, molecular modeling techniques have been used to understand the molecular mechanism and the rate of electron transfer. Mehra and Kepp, in 2019, have shown that the electron transfer promotes the energetical reorganization over a protein and contributing to the catalytic turnover (Mehra & Kepp, 2019). The glycosylation is another import topic in laccase, Glazunova et al., in 2019, studied the carbohydrate moieties in from *Steccherinum murashkinskyi* laccase structure, discovering that N-linked glycosylation in crucial catalytic parameters in phenolic substrates oxidation (Glazunova et al., 2019). Orlikowska et al., 2018, studied glycosylation modeling of *Pycnoporus sanguineus* laccases identified that ASN354 N-glycosylation site, close to the substrate-binding pocket which is particularly significant to the efficient catalysis in laccases (Orlikowska et al., 2018). Gabdulkhakov et al., in 2019 studied a bacterial two-domains laccase with a trinuclear copper center present in three-domain laccases and probe in-silico mutations HIS165PHE, HIS165ALA, ILE172ALA and ILE172PHE, founding that e HIS165ALA variant is more active than the wild type *Streptomyces griseoflavus* laccase (Gabdulkhakov et al., 2019).

Mehra et al., in 2018, used laccase homology model to rational design of new laccases with biotechnological applications. In addition, they confirm that the ASP206 and HIS458 residues are important in substrates binding to *Trametes versicolor* laccase. Moreover, Herrera-Zuñiga et al., in 2018, used homology modeling techniques to study laccase structure evolution from fungi, bacteria, insects, and plants using more than a 100 laccase structures to understand mainly differences in the active site (Rubén Díaz et al., 2018). Structure-function techniques were used by Glazunova et al., in 2018, to comprehend oxidized - reduced states and compared the position and coordination of copper ions in *Antrodiella faginea* and *Steccherinum murashkinskyi* laccases to classify the structure of T1 copper (Glazunova et al., 2018). In other similar studies, *Thermus thermophilus* laccase the geometry formed by the primary coordination ligands is tuned by the E356, E456, D106, and D423 or most commonly second-sphere residues via H-bonding networks (Liu et al., 2018). Kumar et al., in 2017, studied by molecular modeling four isozymes of *Ganoderma lucidum* MDU-7 and *Ganoderma sp kk-02* laccases with different substrates, they reported significant differences in its binding affinity (Kumar et al., 2017). In a similar work, laccase isoenzymes of *Trametes versicolor* studied provide a structural insight into mycotoxins affinity to laccase, this work prove that protein interaction-engineering strategies in laccase sequence help to module laccase-toxin affinities (Dellafiora et al., 2017). Similarly, to elucidate molecular basis of thermostable laccase activity of *Myceliophthora thermophila*, Ernst et al., in 2018, focused in the residues located on the surface affecting laccase catalysis. This laccase has different topologies in its active site and its N-glycosylated motifs are located in the interfacial regions when comparing with other laccases, in particular the ASN88 and ASN210 sites are evolutionarily arranged as an integral part of the laccase surface of *Myceliophthora thermophila* (Ernst et al., 2018).

3. LACCASE DOCKING

In this section, it is presented a compilation of studies carried out in the last years in laccase-ligand interaction strategies to elucidate substrates, mediators or inhibitors interactions, as well as docking studies focused on the biotransformation of anthropogenic molecules harmful to the environment. Laccase-pesticides interaction studied by Rudakiya et al., in 2020 revels the importance of H83, H320, A95, V384, and P366 residues near to the active site, these residues promote a higher binding efficacy on laccase active site to profenofos, chlorpyrifos and thiophanatmethyl pesticide efficiently (Rudakiya et al., 2020). Conceição et al., 2019, studied the interaction between cinnamic acid and *Trametes versicolor* laccase by docking methodology, showing analogs i.e., phenol ring substituents that can shift the active site orientation and conformation (Conceição et al., 2019). In another work carry out by Mo et al., in 2018, with *Trametes versicolor* laccase, sixteen environmental toxics nonylphenol and octylphenol lineal isomers proves that the affinity interaction in this system is a function of the chemical substituent in the ligand which follows as stereochemical position: para < meta < ortho (Mo et al., 2018). In the same context laccase ability to degraded herbicide was performed by docking studies between *Ceriporiopsis subvermispora* laccase and diuron, it was found that LYS457 residue must be necessarily protonated for correct catalysis of diuron (Vieira et al., 2015). In order to find the best catalytic capacity for the enzyme laccase, Behbahani et al., 2019, conducted comparative studies of phenolic and non-phenolic substrates in laccases of fungal and those from bacterial origin using Autodock software. Results showed that the interaction complex energies to protein-ligand are stronger in fungal laccases in comparison with bacterial ones, suggesting a better association and physicochemical efficacy in fungal enzymes (Behbahani et al., 2020). To probe the pH stabilization Novoa et al., 2019, performed molecular docking in two mutants found by directed evolution mutagenesis in *Melanocarpus albomyces* fungal laccase. Their results demonstrated that the mutations localize in the surface LEU365GLU and LEU513MET improve and stabilize the structure of the protein, in addition, the variant with two mutations LEU365GLU-LEU513MET acted in a synergic way (Novoa et al., 2019). In food research, laccase interactions have been studied in contact with two minor polyphenols of black tea, these functional food materials have potential as antiobesity and antiperiodontal drugs. In this work Itoh et al., 2017, probe epitheaflagallin and epitheaflagallin 3-O-gallate molecules with *Trametes sp* M120 laccase and the energy affinities by docking. Results, suggest that laccase interact with both compounds the extract from the black tea treatment and these interactions can be beneficial for human health (Itoh et al., 2017). On the other hand, the inhibitory binding research in *Myceliophthora thermophila* laccase in 2019, confirm the importance of hydrogen bonds and the hydrophobic force to maintain the binding sites. The results indicate that it is important to carry out an integrated study of enzyme kinetics and molecular simulations, in order to establish quantitative and qualitative concepts of the Hofmeister series. In this context by docking electrostatic interactions, it was concluded that LEU429 can influence stereo orientations to perturb electrostatics in His508 at contact with ABTS (Jianliang et al., 2017). In similar work, Martínez-Sotres et al., 2015, performed molecular docking to know the activity inhibition in lignin degradation of *Trametes versicolor* by medicarpin molecule, the results demonstrated to be favorable for developing a wood-preserving analog of medicarpin (Martínez-Sotres et al. 2015).

4. MOLECULAR DYNAMICS SIMULATIONS AND QUANTUM APPROACHES OF LACCASE

The purpose of Molecular Dynamics (MD) simulations is to analyze thermodynamic the kinetic properties of ligand-binding events. It is an effective tool to model biochemical processes, evaluate different physicochemical parameters in protein and break with the static view provided by docking approaches. In this section the most recent molecular dynamics researches are discussed. In the decolorization process of malachite green dye by two *Setosphaeria turcica* laccase isozymes performed by MD, the glycosylation model and dynamics allowed to demonstrate an incorrect location of N-glycosylation sites (Asn-X-Ser/Thr), which is unfavorable for the correct catalysis of laccase, especially those N-glycosylation sites focused on Asn97 residue (Liu et al., 2019). An *Escherichia coli* a special bacterial laccase of six structural domains recently was reported by Schwaneberg et al., in 2020. They suggested that the fifth copper plays a crucial role in the catalytic reactions, eleven variants were built to understand that the electrons transfer is governs by the fifth copper. A similar research was done by Liu et al., in 2018, using surfactants which can enhance the biodegradation of phenolic wastewater by laccase, however the molecular mechanism is still unclear.

In that work the binding mechanisms of phenol with Laccase from *Trametes versicolor* was investigated in the presence or absence of triton X-100 and rhamnolipid by molecular docking simulations and molecular dynamics, results indicate that phenol interacts with the active laccase site via hydrogen bonds and van der Waals interactions in aqueous solutions. The presence of triton X-100 or rhamnolipid results in significant changes in laccase enzymatic conformations, the hydrophobic parts of surfactants contact with the outside surface of laccase. These changes lead to the decrease of binding energy between phenol and laccase. The migration behavior of water molecules within hydration shell is also inevitably affected. Therefore, the amphipathic triton X-100 or rhamnolipid may influence the phenol degradation ability of laccase by modulating their interactions and water environment. Finally, when laccase changed the native conformation, phenol degradation in industrial effluents can decrease (Liu et al., 2018). The structural behavior with the same laccase was studied by molecular dynamics simulations in methanol and hexane solvents. This laccases has better stability in aqueous conditions and is unstable in organic solvents due to its high secondary structure of β-barrel architecture and it was found that the forces which stabilize the tertiary structure play a pivotal role in the enzyme tolerance of both polar and non-polar organic solvents (Mohtashami et al., 2019). Laccase polymerization of catechol was performed using various reactors, i.e., water bath, an ultrasonic bath and a high-pressure homogenizer. Total free OH content and the resulting MALDI-TOF spectra of polymers have shown that reactions are preferred in the presence of high-energy environments. Higher conversion yields and polymerization degrees were obtained. These results were supported by MD simulation studies showed a more open enzyme active site in the presence of high-energy environments or high molecular agitation (Su et al., 2018). Bioremediation by *Fusarium culmorum* laccase of contaminated sites with di (2-ethyl hexyl) phthalate, a plasticizer widely used in the manufacture of plastics, was studied by Ahuactzin-Pérez in 2016 to better understand its biodegradation pathway. Ahuactzin-Pérez et al. in 2016 carried out a Quantum Chemical modeling analysis, which identified intermediate reaction compounds and electron flow, a finding that is very significant for understanding the use of laccases in favor of the environment (Ahuactzin-Pérez et al., 2016)

CONCLUSION

Many of laccase enzymes have been engineered, combining rational and computational design with directed evolution, to attain the stability, catalytic efficiency and selectivity properties required for their bioindustrial utilization. Almost all laccases with an *ad-hoc* software and computational strategies reviewed through this chapter, shown that the structure, catalytic activity, resistance and tolerance capabilities can be predicted, improved or changed by means of biophysical and biochemical simulations, reducing the research time in several orders of magnitude while laccase are screening and engineering them *in-silico*. In other words, computational studies may help to improve or change the bioindustrial uses of laccases that still seem limited, focusing mostly in biodegradation applications. Due to the challenges generated in the fields of application using new methodologies in the laccase studies, the research interest in this type of enzymes is far from diminishing. Recently, new application areas for laccases have emerged, that requires computational tools to allow accelerated laccases development to adapt them to new industrial applications.

REFERENCES

Acosta-Urdapilleta, L., Paz, G. A. A. P., Rodríguez, A., Adame, M., Salgado, D., Salgado, J., Villegas Villareal, E. C. (2010). Pycnoporus sanguineus, un hongo con potencial biotecnológico [Pycnoporus sanguineus, a fungus with biotechnological potential]. In *Hacia un Desarrollo Sostenible del Sistema de Producción-Consumo de los Hongos Comestibles y Medicinales en Latinoamérica: Avances y Perspectivas en el Siglo XXI.*

Ahlawat, S., Singh, D., Virdi, J. S., Sharma, K. K. (2019). Molecular modeling and MD- simulation studies: Fast and reliable tool to study the role of low-redox bacterial laccases in the decolorization of various commercial dyes. *Environmental Pollution* 253, 1056-1065. https://doi.org/10.1016/j.envpol.2019.07.083.

Ahuactzin-Pérez, M., Tlecuitl-Beristain, S., García-Dávila, J., González-Pérez, M., Gutiérrez-Ruíz, M. C., Sánchez, C. (2016). Degradation of di(2-ethyl hexyl) phthalate by Fusarium culmorum: Kinetics, enzymatic activities and biodegradation pathway based on quantum chemical modeling pathway based on quantum chemical modeling. *Science of the Total Environment* 566-567, 1186-1193. https://doi.org/10.1016/j.scitotenv.2016.05.169.

Arregui, L., Ayala, M., Gómez-Gil, X., Gutiérrez-Soto, G., Hernández-Luna, C. E., Herrera De Los Santos, M., Valdez-Cruz, N. A. (2019). Laccases: structure, function, and potential application in water bioremediation. *Microbial Cell Factories* 18, 200 https://doi.org/10.1186/s12934-019-1248-0.

Avelar, M., Pastor, N., Ramirez-Ramirez, J., Ayala, M. (2018). Replacement of oxidizable residues predicted by QM-MM simulation of a fungal laccase generates variants with higher operational stability. *Journal of Inorganic Biochemistry* 178, 125-133. https://doi.org/10.1016/j.jinorgbio.2017.10.007.

Awasthi, M., Jaiswal, N., Singh, S., Pandey, V. P., Dwivedi, U. N. (2015). Molecular docking and dynamics simulation analyses unraveling the differential enzymatic catalysis by plant and fungal laccases with respect to lignin biosynthesis and degradation. *Journal of Biomolecular Structure and Dynamics* 33(9), 1-52 https://doi.org/10.1080/07391102.2014.975282.

Ayla, S., Kallubai, M., Pallipati, S. D., Narasimha, G. (2019). Enzymatic Textile Dyes Decolorization by *In vitro* and *In silico* Studies. *Recent Patents on Biotechnology* 13(4), 268-276. https://doi.org/10.2174/1872208313666190625123847.

Behbahani, M., Nosrati, M., Moradi, M., Mohabatkar, H. (2020). Using Chou's General Pseudo Amino Acid Composition to Classify Laccases from Bacterial and Fungal Sources via Chou's Five-Step Rule. *Applied Biochemistry and Biotechnology* 190, 1035-1048. https://doi.org/10.1007/s12010-019-03141-8.

Benson, D., Boguski, M., Lipman, D. J., Ostell, J. (1990). The National Center for Biotechnology Information. *Genomics* 6, 389-391. https://doi.org/10.1016/0888-7543(90)90583-G.

Chaurasia, K., P., L. Bharati, S., Sarma, C. (2016). Laccases in Pharmaceutical Chemistry: A Comprehensive Appraisal. *Mini-Reviews in Organic Chemistry* 13(6), 430-451. https://doi.org/10.2174/1570193x13666161019124854.

Christensen, N. J., Kepp, K. P. (2013). Stability Mechanisms of a Thermophilic Laccase Probed by Molecular Dynamics. *PLoS ONE* 8(4), e61985 https://doi.org/10.1371/journal.pone.0061985.

Conceição, J. C. S., Dias, H. J., Peralva, C. M. S., Crotti, A. E. M., da Rocha Pita, S. S., de Oliveira Silva, E. (2019). Phenolic Compound Biotransformation by Trametes versicolor ATCC 200801 and Molecular Docking Studies. *Applied Biochemistry and Biotechnology* 190, 1498-1511 https://doi.org/10.1007/s12010-019-03191-y 506.

de Salas, F., Cañadas, R., Santiago, G., Virseda-Jerez, A., Vind, J., Gentili, P., Camarero, S. (2019). Structural and biochemical insights into an engineered high-redox potential laccase overproduced in Aspergillus. *International Journal of Biological Macromolecules* 141, 855-867. https://doi.org/10.1016/j.ijbiomac.2019.09.05.

Dellafiora, L., Galaverna, G., Reverberi, M., Dall'Asta, C. (2017). Degradation of aflatoxins by means of laccases from trametes versicolor: An *in-silico* insight. *Toxins* 9, 17. https://doi.org/10.3390/toxins9010017.

Desai, S. S., & Nityanand, C. (2011). Microbial Laccases and their Applications: A Review. *Asian Journal of Biotechnology* 3(2), 98-124. https://doi.org/10.3923/ajbkr.2011.98.124.

Devi, P., Kandasamy, S., Uthandi, S. (2016). Laccase producing Streptomyces bikiniensis CSC12 isolated from compost. *Journal of Microbiology, Biotechnology and Food Sciences* 6(2), 794-798. https://doi.org/10.15414/jmbfs.2016.6.2.794-798.

Dhakar, K., Pandey, A. (2013). Laccase production from a temperature and pH tolerant fungal strain of Trametes hirsuta (MTCC 11397). *Enzyme Research* 2013, e(9). https://doi.org/10.1155/2013/869062.

Díaz, R., Díaz-Godínez, G., Anducho-Reyes, M. A., Mercado-Flores, Y., Herrera-Zúñiga, L. D. (2018). *In silico* design of laccase thermostable mutants from lacc 6 of pleurotusm ostreatus. *Frontiers in Microbiology* 9, 2743. https://doi.org/10.3389/fmicb.2018.02743.

Díaz, Rubén, Yuridia, M. F., Díaz-Godínez, G., Herrera-Zúñiga, L. D., álvarez-Cervantes, J., Anducho-Reyes, M. A. (2018). *In silico* generation of laccase mutants from lacc 6 of Pleurotus ostreatus and bacterial enzymes. *BioResources* 13(4), 8113-8131. https://doi.org/10.15376/biores.13.4.8113-8131.

Ducros, V., Brzozowski, A. M., Wilson, K. S., Brown, S. H., Østergaard, P., Schneider, P Davies, G. J. (1998). Crystal structure of the type-2 Cu depleted laccase from Coprinus cinereus at 2.2 Å resolution. *Nature Structural Biology* 5, 310-316. https://doi.org/10.1038/nsb0498-310.

Eggert, C. (1997). Laccase-catalyzed formation of cinnabarinic acid is responsible for antibacterial activity of Pycnoporus cinnabarinus. *Microbiological Research* 152(3), 315–318. https://doi.org/10.1016/s0944-5013(97)80046-8.

Ernst, H. A., Jørgensen, L. J., Bukh, C., Piontek, K., Plattner, D. A., Østergaard, L. H., Bjerrum, M. J. (2018). A comparative structural analysis of the surface properties of asco-laccases. *PLoS ONE* 13(11), e0206589. https://doi.org/10.1371/journal.pone.0206589.

Ferrario, V., Chernykh, A., Fiorindo, F., Kolomytseva, M., Sinigoi, L., Myasoedova, N., Gardossi, L. (2015). Investigating the Role of Conformational Effects on Laccase Stability and Hyperactivation under Stress Conditions. *Chem Bio Chem* 16, 2365-2372. https://doi.org/10.1002/cbic.201500339 556.

Gabdulkhakov, A., Kolyadenko, I., Kostareva, O., Mikhaylina, A., Oliveira, P., Tamagnini, P., Tishchenko, S. (2019). Investigations of accessibility of T2/T3 copper center of two-domain laccase from streptomyces griseoflavus ac-993. *International Journal of Molecular Sciences* 20, 3184. https://doi.org/10.3390/ijms20133184.

Garcia, K. E., Babanova, S., Scheffler, W., Hans, M., Baker, D., Atanassov, P., Banta, S. (2016). Designed protein aggregates entrapping carbon nanotubes for bioelectrochemical oxygen reduction. *Biotechnology and Bioengineering* 113(11), 2321-2327. https://doi.org/10.1002/bit.25996.

Glazunova, O. A., Moiseenko, K. V., Kamenihina, I. A., Isaykina, T. U., Yaropolov, A. I., Fedorova, T. V. (2019). Laccases with variable properties from different strains of steccherinum ochraceum: Does glycosylation matter? *International Journal of Molecular Sciences* 20, 2008. https://doi.org/10.3390/ijms20082008.

Glazunova, O. A., Polyakov, K. M., Moiseenko, K. V., Kurzeev, S. A., Fedorova, T. V. (2018). Structure-function study of two new middle-redox potential laccases from basidiomycetes *Antrodiella faginea* and *Steccherinum murashkinskyi*. *International Journal of Biological Macromolecules* 118(A), 406-418. https://doi.org/10.1016/j.ijbiomac.2018.06.038.

Herrera-Zúñiga, L. D., Millán-Pacheco, C., Viniegra-González, G., Villegas, E., Arregui, L., Rojo-Domínguez, A. (2019). Molecular dynamics on laccase from *Trametes versicolor* to examine thermal stability induced by salt bridges. *Chemical Physics*, 517, 253-264. https://doi.org/10.1016/j.chemphys.2018.10.019 580.

Hildén, K., Hakala, T. K., Lundell, T. (2009). Thermotolerant and thermostable laccases. *Biotechnology Letters* 31, 1117. https://doi.org/10.1007/s10529-009-9998-0.

Hongyan, L., Zexiong, Z., Shiwei, X., He, X., Yinian, Z., Haiyun, L., Zhongsheng, Y. (2019). Study on transformation and degradation of bisphenol A by *Trametes versicolor* laccase and simulation of molecular docking. *Chemosphere* 224, 743-750. https://doi.org/10.1016/j.chemosphere.2019.02.143.

Itoh, N., Kurokawa, J., Isogai, Y., Ogasawara, M., Matsunaga, T., Okubo, T., Katsube, Y. (2017). Functional Characterization of Epitheaflagallin 3-O-Gallate Generated in Laccase-Treated Green Tea Extracts in the Presence of Gallic Acid. *Journal of Agricultural and Food Chemistry* 65(48), 10473-10481. https://doi.org/10.1021/acs.jafc.7b04208.

Janusz, G., Pawlik, A., Świderska-Burek, U., Polak, J., Sulej, J., Jarosz-Wilkołazka, A., Paszczyński, A. (2020). Laccase properties, physiological functions, and evolution. *International Journal of Molecular Sciences* 21, 966. https://doi.org/10.3390/ijms21030966.

Jianliang, S., Liu, H., Yang, W., Chen, S., Fu, S. (2017). Molecular mechanisms underlying inhibitory binding of alkylimidazolium ionic liquids to laccase. *Molecules* 22, 1353. https://doi.org/10.3390/molecules22081353.

Jones, S. M., Solomon, E. I. (2015). Electron transfer and reaction mechanism of laccases. *Cellular and Molecular Life Sciences* 72(5), 869-883. https://doi.org/10.1007/s00018-014-1826-6.

Julio, A. H. F., Gigliolli, A. A. S., Cardoso, K. A. K., Drosdoski, S. D., Kulza, R. A., Seixas, F. A. V., ... Lapenta, A. S. (2017). Multiple resistance to pirimiphos-methyl and bifenthrin in Tribolium castaneum involves the activity of lipases, esterases, and laccase2. Comparative Biochemistry and Physiology Part - C: *Toxicology and Pharmacology* 195, 27-43. https://doi.org/10.1016/j.cbpc.2017.01.011.

Kameshwar, A. K. S., Barber, R., Qin, W. (2018). Comparative modeling and molecular docking analysis of white, brown and soft rot fungal laccases using lignin model compounds for understanding the structural and functional properties of laccases. *Journal of Molecular Graphics and Modelling* 79, 15-26. https://doi.org/10.1016/j.jmgm.2017.10.019.

Kepp, K. P. (2015). Halide binding and inhibition of laccase copper clusters: The role of reorganization energy. *Inorganic Chemistry*. https://doi.org/10.1021/ic5021466.

Kim, H., Sharma, S. K., Schaefer, A. W., Solomon, E. I., Karlin, K. D. (2019). Heme-Cu Binucleating Ligand Supports Heme/O2 and FeII-CuI/O2 Reactivity Providing High- And Low-Spin FeIII-Peroxo-CuII Complexes. *Inorganic Chemistry* 58(22), 15423-15432. https://doi.org/10.1021/acs.inorgchem.9b02521.

Komori, H., Miyazaki, K., Higuchi, Y. (2009). X-ray structure of a two-domain type laccase: A missing link in the evolution of multi-copper proteins. *FEBS Letters* 583(7), 1189-1195. https://doi.org/10.1016/j.febslet.2009.03.008.

Kudanga, T., Le Roes-Hill, M. (2014). Laccase applications in biofuels production: Current status and future prospects. *Applied Microbiology and Biotechnology* 98, 6525-6542. https://doi.org/10.1007/s00253-014-5810-8.

Kumar, A., Singh, D., Sharma, K. K., Arora, S., Singh, A. K., Gill, S. S., Singhal, B. (2017). Gel-based purification and biochemical study of laccase isozymes from Ganoderma sp. and its role in enhanced cotton callogenesis. *Frontiers in Microbiology* 8, 674. https://doi.org/10.3389/fmicb.2017.00674.

Laskowski, R. A., Jabłońska, J., Pravda, L., Vařeková, R. S., Thornton, J. M. (2018). PDBsum: Structural summaries of PDB entries. *Protein Science* 27(19), 129-134. https://doi.org/10.1002/pro.3289.

Liu, H., Zhu, Y., Yang, X., & Lin, Y. (2018). Four second-sphere residues of Thermus thermophilus SG0.5JP17-16 laccase tune the catalysis by hydrogen-bonding networks. *Applied Microbiology and Biotechnology* 102, 4049-4061. https://doi.org/10.1007/s00253-018-8875-y.

Liu, N., Shen, S., Jia, H., Yang, B., Guo, X., Si, H., Dong, J. (2019). Heterologous expression of Stlac2, a laccase isozyme of Setosphearia turcica, and the ability of decolorization of malachite green. *International Journal of Biological Macromolecules*. https://doi.org/10.1016/j.ijbiomac.2019.07.029.

Liu, Y., Liu, Z., Zeng, G., Chen, M., Jiang, Y., Shao, B., Liu, Y. (2018). Effect of surfactants on the interaction of phenol with laccase: Molecular docking and molecular dynamics simulation studies. *Journal of Hazardous Materials* 138, 21-28. https://doi.org/10.1016/j.jhazmat.2018.05.042.

Lu, J., Shao, J., Liu, H., Wang, Z., Huang, Q. (2015). Formation of Halogenated Polyaromatic Compounds by Laccase Catalyzed Transformation of Halophenols. *Environmental Science and Technology* 49(14), 8550-8557. https://doi.org/10.1021/acs.est.5b02399.

Lucas, M. F., Monza, E., Jørgensen, L. J., Ernst, H. A., Piontek, K., Bjerrum, M. J., Guallar, V. (2017). Simulating Substrate Recognition and Oxidation in Laccases: From Description to Design. *Journal of Chemical Theory and Computation* 13(3), 1462-1467. https://doi.org/10.1021/acs.jctc.6b01158.

Machczynski, M. C., Vijgenboom, E., Samyn, B., Canters, G. W. (2004). Characterization of SLAC: A small laccase from Streptomyces coelicolor with unprecedented activity. *Protein Science* 13, 2388-2397. https://doi.org/10.1110/ps.04759104.

Madhavan, S., Krause, K., Jung, E. M., Kothe, E. (2014). Differential regulation of multi-copper oxidases in Schizophyllum commune during sexual development. *Mycological Progress* 13, 1009. https://doi.org/10.1007/s11557-014-1009-8.

Maniak, H., Talma, M., Matyja, K., Trusek, A., Giurg, M. (2020). Synthesis and Structure-Activity Relationship Studies of Hydrazide-Hydrazones as Inhibitors of Laccase from Trametes versicolor. *Molecules* 25, 1255. https://doi.org/10.3390/molecules25051255.

Martínez-Sotres, C., Rutiaga-Quiñones, J. G., Herrera-Bucio, R., Gallo, M., López- Albarrán, P. (2015). Molecular docking insights into the inhibition of laccase activity by medicarpin. *Wood Science and Technology* 49(4). https://doi.org/10.1007/s00226-015-0734-8.

Mate, D. M., Alcalde, M. (2017). Laccase: a multi-purpose biocatalyst at the forefront of biotechnology. *Microbial Biotechnology* 10(6), 1457-1467. https://doi.org/10.1111/1751-7915.12422.

Mehra, R., Kepp, K. P. (2019). Contribution of substrate reorganization energies of electron transfer to laccase activity. *Physical Chemistry Chemical Physics* 21, 15805-15814. https://doi.org/10.1039/c9cp01012b.

Mehra, R., Muschiol, J., Meyer, A. S., Kepp, K. P. (2018). A structural-chemical explanation of fungal laccase activity. *Scientific Reports* 8, 17285. https://doi.org/10.1038/s41598-697 018-35633-8.

Mo, D., Zeng, G., Yuan, X., Chen, M., Hu, L., Li, H., Cheng, M. (2018). Molecular docking simulation on the interactions of laccase from Trametes versicolor with nonylphenol and octylphenol isomers.

Bioprocess and Biosystems Engineering 41, 331-343. https://doi.org/10.1007/s00449-017-1866-z.

Mohtashami, M., Fooladi, J., Haddad-Mashadrizeh, A., Housaindokht, M. R., Monhemi, H. (2019). Molecular mechanism of enzyme tolerance against organic solvents: Insights from molecular dynamics simulation. *International Journal of Biological Macromolecules* 122, 914-923. https://doi.org/10.1016/j.ijbiomac.2018.10.172.

Mojsov, K. (2014). Biotechnological applications of laccases in the textile industry. *Savremene Tehnologije* 3(1), 76-79. https://doi.org/10.5937/savteh1401076m.

Monza, E., Lucas, M. F., Camarero, S., Alejaldre, L. C., Martínez, A. T., Guallar, V. (2015). Insights into laccase engineering from molecular simulations: Toward a binding-focused strategy. *Journal of Physical Chemistry Letters* 6(8), 1447-1453. https://doi.org/10.1021/acs.jpclett.5b00225.

Moreno, A. D., Ibarra, D., Eugenio, M. E., Tomás-Pejó, E. (2020). Laccases as versatile enzymes: from industrial uses to novel applications. *Journal of Chemical Technology and Biotechnology* 95(3), 481-494. https://doi.org/10.1002/jctb.6224.

Novoa, C., Dhoke, G. V., Mate, D. M., Martínez, R., Haarmann, T., Schreiter, M., Schwaneberg, U. (2019). KnowVolution of a Fungal Laccase toward Alkaline pH. *Chem Bio Chem* 3, 1458-1466. https://doi.org/10.1002/cbic.201800807.

Orlikowska, M., de J. Rostro-Alanis, M., Bujacz, A., Hernández-Luna, C., Rubio, R., Parra, R., Bujacz, G. (2018). Structural studies of two thermostable laccases from the white-rot fungus Pycnoporus sanguineus. *International Journal of Biological Macromolecules* 107(B), 1629-1640. https://doi.org/10.1016/j.ijbiomac.2017.10.024.

Osma, J. F., Toca-Herrera, J. L., Rodríguez-Couto, S. (2010). Uses of laccases in the food industry. *Enzyme Research*. https://doi.org/10.4061/2010/918761.

Robert, V., Monza, E., Tarrago, L., Sancho, F., De Falco, A., Schneider, L., Tron, T. (2017). Probing the Surface of a Laccase for Clues towards the Design of Chemo-Enzymatic Catalysts. *Chem Plus Chem* 82(4), 607-614. https://doi.org/10.1002/cplu.201700030.

Rudakiya, D. M., Patel, D. H., Gupte, A. (2020). Exploiting the potential of metal and solvent tolerant laccase from Tricholoma giganteum AGDR1 for the removal of pesticides. *International Journal of Biological Macromolecules* 140, 586-595. https://doi.org/10.1016/j.ijbiomac.2019.12.068.

Ryde, U., Hsiao, Y. W., Rulšek, L., Solomon, E. I. (2007). Identification of the peroxy adduct in multicopper oxidases by a combination of computational chemistry and extended X-ray absorption fine-structure measurements. *Journal of the American Chemical Society* 129(4), 726-727. https://doi.org/10.1021/ja062954g.

Rydén, L. G., & Hunt, L. T. (1993). Evolution of protein complexity: The blue copper- containing oxidases and related proteins. *Journal of Molecular Evolution* 36, 41-66. https://doi.org/10.1007/BF02407305.

Schwaneberg, U., Zhang, L., Cui, H., Dhoke, G. V., Zou, Z., Sauer, D. F., Davari, M. D. (2020). Engineering of Laccase CueO for Improved Electron Transfer in Bioelectrocatalysis by Semi-Rational Design. *Chemistry – A European Journal* 26(22), 4974-4979. https://doi.org/10.1002/chem.201905598.

Singh, D., Rawat, S., Waseem, M., Gupta, S., Lynn, A., Nitin, M., Sharma, K. K. (2016). Molecular modeling and simulation studies of recombinant laccase from Yersinia enterocolitica suggests significant role in the biotransformation of non-steroidal anti- inflammatory drugs. *Biochemical and Biophysical Research Communications* 469(2), 306-312. https://doi.org/10.1016/j.bbrc.2015.11.096.

Singh, G., Arya, S. K. (2019). Utility of laccase in pulp and paper industry: A progressive step towards the green technology. *International Journal of Biological Macromolecules* 134(1), 1070-1084. https://doi.org/10.1016/j.ijbiomac.2019.05.168.

Singh, N. S., Sharma, R., Singh, D. K. (2019). Identification of enzyme(s) capable of degrading endosulfan and endosulfan sulfate using *in silico* techniques. *Enzyme and Microbial Technology* 124, 32-40. https://doi.org/10.1016/j.enzmictec.2019.01.003.

Solomon, E. I., Dooley, D. M., Wang, R. H., Gray, H. B., Cerdonio, M., Mogno, F., Romani, G. L. (1976). Susceptibility Studies of Laccase and Oxyhemocyanin using an Ultrasensitive Magnetometer. Antiferromagnetic Behavior of the Type 3 Copper in Rhus Laccase. *Journal of the American Chemical Society* 98(4), 1029-1031. https://doi.org/10.1021/ja00420a035.

Srinivasan, S., Sadasivam, S. K., Gunalan, S., Shanmugam, G., Kothandan, G. (2019). Application of docking and active site analysis for enzyme linked biodegradation of textile dyes. *Environmental Pollution* 248, 599-608. https://doi.org/10.1016/j.envpol.2019.02.080.

Stanzione, I., Pezzella, C., Giardina, P., Sannia, G., Piscitelli, A. (2020). Beyond natural laccases: extension of their potential applications by protein engineering. *Applied Microbiology and Biotechnology* 104, 915-924. https://doi.org/10.1007/s00253-019-10147-z.

Su, J., Castro, T. G., Noro, J., Fu, J., Wang, Q., Silva, C., Cavaco-Paulo, A. (2018). The effect of high-energy environments on the structure of laccase-polymerized poly(catechol). *Ultrasonics Sonochemistry* 48, 275-280. https://doi.org/10.1016/j.ultsonch.2018.05.033.

Toledo-Núñez, C., López-Cruz, J. I., Hernández-Arana, A. (2012). Thermal denaturation of a blue-copper laccase: Formation of a compact denatured state with residual structure linked to pH changes in the region of histidine protonation. *Biophysical Chemistry* 167, 36-42. https://doi.org/10.1016/j.bpc.2012.04.004.

Tollin, G., Meyer, T. E., Cheddar, G., Getzoff, E. D., Cusanovich, M. A. (1986). Transient Kinetics of Reduction of Blue Copper Proteins by Free Flavin and Flavodoxin Semiquinones. *Biochemistry* 25(11), 3363-3370. https://doi.org/10.1021/bi00359a041.

Valderrama, B., Oliver, P., Medrano-Soto, A., Vazquez-Duhalt, R. (2003). Evolutionary and structural diversity of fungal laccases. *Antonie van Leeuwenhoek* 84, 289-299. https://doi.org/10.1023/A:1026070122451.

Vasin, A., Klotchenko, S., Puchkova, L. (2013). Phylogenetic Analysis of Six-Domain Multi-Copper Blue Proteins. *PLoS Currents*. https://doi.org/10.1371/currents.tol.574bcb0f133fe52835911abc4e296141.

Vieira, A. C., Marschalk, C., Biavatti, D. C., Lorscheider, C. A., Peralta, R. M., Augusto, F., Seixas, V. (2015). Modeling Based Structural Insights into Biodegradation of the Herbicide Diuron by Laccase-1 from Ceriporiopsis subvermispora. *Bioinformation* 11(5), 224-228. https://doi.org/10.6026/97320630011224.

Viswanath, B., Rajesh, B., Janardhan, A., Kumar, A. P., Narasimha, G. (2014). Fungal laccases and their applications in bioremediation. *Enzyme Research*, 2014, 163242. https://doi.org/10.1155/2014/163242.

Wang, Y. F., Xu, H., Feng, L., Shen, X. F., Wang, C., Huo, X. K., Deng, S. (2019). Oxidative coupling of coumarins catalyzed by laccase. *International Journal of Biological Macromolecules* 135, 1028-1033. https://doi.org/10.1016/j.ijbiomac.2019.05.215.

Zaccaria, M., Dawson, W., Cristiglio, V., Reverberi, M., Ratcliff, L. E., Nakajima, T., Momeni, B. (2020). Designing a bioremediator: mechanistic models guide cellular and molecular specialization. *Current Opinion in Biotechnology* 62, 98-105. https://doi.org/10.1016/j.copbio.2019.09.006.

In: Molecular Basis of Specific Mechanism ... ISBN: 978-1-53618-751-9
Editors: Marcos López-Pérez et al.   © 2020 Nova Science Publishers, Inc.

Chapter 6

HOST DIET AND HOST DERIVED GLYCANS AS PRIMARY DRIVERS OF MICROBIAL GUT ADAPTATION

Rina González-Cervantes[*], *Félix Aguirre-Garrido*
and Marcos López-Pérez
Environmental Science Department. Autonomous Metropolitan University, (Lerma), Mexico City, Mexico

ABSTRACT

The composition of the food diet is intrinsically variable, however the presence of certain compounds such as glycans exerts an important action with respect to different adaptive processes of bacterial species that are part of the intestinal microbiota. In this chapter we review the main works that delve into the molecular bases of the adaptive mechanisms of these microorganisms to intestinal conditions, making special mention of the mechanisms of cellular internalization and signaling that enable the optimization of adaptive processes under symbiotic physiological conditions.

Keywords: host diet, glycans, microbial adaptation

---

[*] Corresponding Author's Email: rgonzalez@correo.ler.uam.mx Environmental Science Department. Autonomous Metropolitan University, (Lerma), Mexico City, Mexico.

1. Introduction

Microorganisms associated with humans constitute huge communities of cells, even ten times more than human cells, it has been estimated that these microbial communities carry information hundreds of times greater than the genes encoded in the human genome (Qin, J et al., 2010). Most of these microorganisms have the intestine as an ecological niche and have been shown to have great relevance in terms of their impact on human physiology and nutrition. On the other hand, these intestinal bacteria perform a basic function in nutrition, such as the capture of energy from nutrients, the stimulation of the immune system as well as highly relevant ecological interactions such as competition phenomena with other microorganisms with pathogenic potential. In this sense, Qin, J. et al. 2010 deepened through a metagenomic analysis based on stool samples from European people (124), the results indicated that 99.1% of the genes obtained in the collection were of bacterial origin, on the other part of determined that only 0.1% corresponded to genes belonging to the *Archaea* domain, other eukaryotic organisms and viral genes. *Bacteroidetes* and *Firmicutes* showed high levels of prevalence. On the other hand, some of the most represented groups were members of the groups *Eubacterium / Bacteroidetes* and *Dorea / lactobacilli / bifidobacteria, proteobacteria* and *streptococci* and *Ruminococcus* (Rinninella et al., 2019). These communities have an interesting characteristic regarding their metabolism and that is that they are highly populated by bacteria with anaerobic metabolism (Thursby & Juge 2017). An element that must be emphasized refers to the process of acquisition and development of the microbiota, where it has been theorized that it begins from birth (Yang et al., 2016), in this sense it has been determined that it can be affected even by the way of delivery of infants (vaginally or by caesarean section). Age is one of the determining parameters to explain the changes of the different dominant bacterial groups. In relation to this same transition, the effect of antibiotic therapy (Bhalodi et al., 2019) as well as diet change has been deepened. Changes of gut microbiota may be associated with a state of health or disease in the human body (Thursby E and N. Juge 2017) Table 1. Besides, host genotype, immune status, and health state can also affect the composition of the gut microbiota, as well as environmental factors derived of the intrinsic characteristics of the gut as an ecosystem such as the presence of a radial oxygen gradient which segregates the gut microbiota radially along the longitudinal axis of the gut (Albenberg, L et al., 2014) and distribution of the tissue-associated mucus (Hai et al., 2015). Though all of the above-mentioned factors affect microbiota in the intestinal tract, the primary driver, in not only microbial community structure but also in microbial gene expression, appears to be the composition and intake levels of host diet (David et al., 2014, Shoaie et al., 2015).

Table 1. Changes of gut microbiota that may be associated with a state of health or disease

Health	Microbial activity or products	Disease
Nutrients and energy supplementation	SCFA production, vitamins synthesis, influence in gut hormones production and energy expenditure	Obesity and metabolic syndrome (Saaed et. al., 2015, Everard et. al., 2013)
Pathogens inhibition	Bacteriocins production, decreased intestinal pH, physical barrier	Potential infection (Harris et. al., 2017)
Immune system development, training, and modulation.	Inflammatory and anti-inflammatory signals balance and production	Allergies IBD Chronic inflammation (Kelly and Mulder 2012)
Mental Health	Neurotransmitters and their precursors production	Anxiety, depression Neurodevelopment (Savignac et al., 2013)

The coevolutionary process over millennia of years between the host and these microorganisms resulted in a symbiotic relationship of the mutualistic type where a feedback is established between beneficial effects caused by the microorganisms in the physiology of the host, and in turn, a supply of nutrients and an ecological niche conducive to bacteria is provided by the host. One of the most relevant effects provided by intestinal microbiota is the strengthening of the intestinal epithelial barrier, homeostatic regulation as well as an adequate immunological response avoiding the colonization of microorganisms with pathogenic potential. In this sense, it should be noted that most of the species present in the microbiota are commensal or mutualistic (Pickard et al., 2017).

2. Carbohydrates in the Intestine as the Main Nutrient That Determines Bacteria Colonization and Adaptation

The environmental conditions in the intestine are due to the homeostatic balance that is established between the physiology of the community and the physiology of the host. The host conditions the values of the parameters of the intestine in various ways. For example, it has been determined that the immune dynamics of the host can act as a selection pressure to limit or expand different niches (Yamada et al., 2019). However, it has been theorized that the mosaic of niches present in the intestinal environment is conditioned both qualitatively and quantitatively by the nutrients present due to the effect of the host's diet, as well as by the presence of different secretions that it discharges into the intestine (Pereira and Berry, 2017). It is relevant to emphasize that nutrient niche partitioning is a prevalent phenomenon in the human gut microbiota (Plichta *et al.*, 2016). In this sense, it has been established that the diversity of nutrients as well as the compounds derived from their interactions has an effect on the colonizing potential of certain species in the microbiota, as published in relevant studies (Freter et al., 1983).

On the other hand, from these same studies it was inferred that it is the molecular mechanisms and processes that make it possible to use one or more limiting nutrients more efficiently than their competitors in the intestinal environment, the fundamental variable that ultimately results in a difference of the adaptation and persistence of some species with respect to others. Considering this scenario as the fundamental variable to explain the adaptive differences between species, it is pertinent to add that the nutrient flows in the intestine fluctuate both qualitatively and quantitatively in space and time.

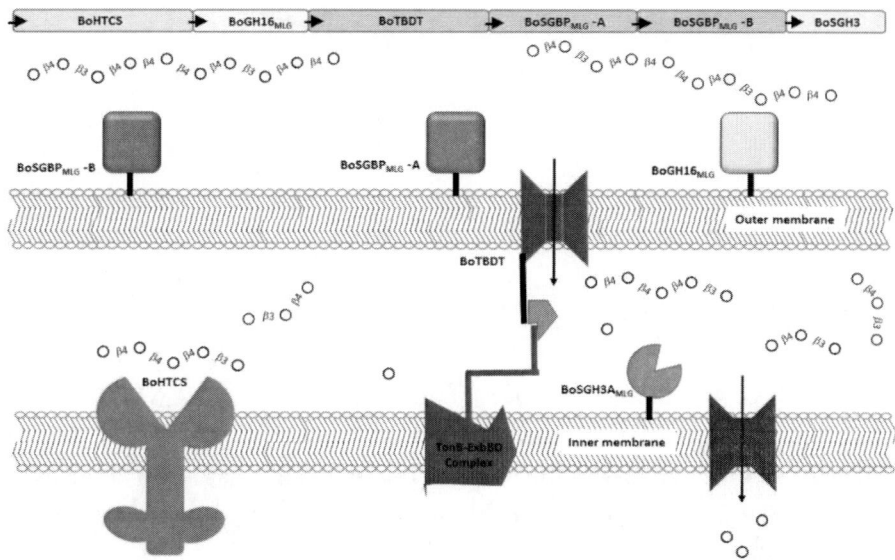

Figure 1. Model of Mixed-Linkage β-Glucan Saccharification by the Concerted Action of the MLGUL Machinery. (Image modified by Tamura et al., 2017).

It is therefore important to refer to the term metabolic flexibility, which can be understood from two perspectives, a) the variations in the mechanisms of expression regulation, that is, bacteria that achieve adaptation along a gradient of possibilities that are a function of the intensity with which a gene or group of genes are expressed in order to take advantage of a nutrient, situation that has already been observed in other studies (Weston et al., 2017), and b), the acquisition of new information under this environment could be by two mechanisms, i) mutation, as has already been published (Garud et al., 2019) and ii) through horizontal gene transfer mechanisms, which has also been published in several relevant studies (Lerner et al., 2017). An example well described in the bibliography that delves into the molecular bases of bacterial adequacy, is related to the colonization processes associated with the use of certain carbohydrates, refers to the mistaken use of β glucans in a group very present in the human microbiota like *Bacteroides* group. (Figure 1)

3. RELEVANCE OF OMICS TECHNIQUES IN THE ANALYSIS OF ADAPTIVE PROCESSES IN THE INTESTINAL MICROBIOTA

To identify the basic functions encoded in the microbiota metagenome, several studies were developed that focused on the articulation of orthologous groups, which were determined to be shared by various individuals under study. Of the known fraction, approximately 5% corresponded to genes associated with bacteriophage proteins, which is indicative of the great relevance of the presence of these viruses in the homeostatic regulation of the microbiota (Quin J, et al., 2010). On the other hand, the most striking secondary metabolism that seems crucial for the minimal metagenome is related, to the biodegradation of complex sugars and glycans harvested from the host's diet and / or the intestinal lining. Examples include degradation and absorption pathways for pectin (and its monomer, rhamnose) and sorbitol, sugars that are ubiquitous in fruits and vegetables, but are not absorbed or poorly absorbed by humans. Since it was discovered that some intestinal microorganisms degrade them both, this capacity seems to have been selected in the intestinal ecosystem as a non-competitive energy source. In addition to these, the ability to ferment, for example, mannose, fructose, cellulose, and sucrose, is also part of the gene pool through the analysis of the minimal metagenome (Qin, J et al., 2010). On the other hand, there are studies that, through the use of metagenomic techniques, have explored the adaptive phenomena in a species with a high prevalence in the human microbiota, *Bacteroides fragilis*; In this sense, it was shown that several genes involved in the use of polysaccharides were monitored under a parallel evolution process that were modified to achieve the adaptation of these species. The availability of diet-derived nutrients changes according to the diet of the host, whereas host-derived nutrients, such as mucin, are constantly available. Therefore, the capacity to use host-derived glycans can confer a competitive advantage to the bacteria residing in the gut (Corfield 2018). Mucus consists mainly of the heavily glycosylated Muc2 and serves several purposes. It acts as a lubricant for food passing over membranes, as a selective barrier to allow the passage of nutrients to the epithelial cells, and as a protective system against mechanical damage or harmful substances such as pathogens and toxins (Corfield & Shukla 2003). Regarding the commensal microbiota, the mucus layer offers the bacteria substrates for growth, adhesion, and protection. Mucins, the main components of the mucus layer, are large glycoproteins secreted by goblet cells in the epithelium and consist of protein backbones decorated with a variety of carbohydrate chains (Lillehoj et al., 2013). The main sugar monomers present in mucin are fucose, galactose, *N*-acetylgalactosamine (GalNAc), and *N*-acetylglucosamine (GlcNAc); in some cases, they are modified with sialic acid (*N*-acetylneuraminic acid) and sulfate. These glycans are attached to the protein backbone via *O*-glycosidic linkages to serine and threonine residues (Ottman et al., 2017).

4. CARBORHYDRATES DERIVED FROM PLANTS INCLUDED IN THE DIET. TRANSPORT ENZYMES AND METABOLISM

In the diet a good part of the carbohydrates that are consumed are of plant origin, which is important from the point of view of their chemical structure and therefore the implications that this has in order to delve into the molecular bases of the mechanisms of degradation. In this sense, one of the most relevant species in which these aspects have been studied are pertinent to the genus Bacteroides. *B. thetaiotaomicron* is a glycophile that can break down a broad array of dietary polysaccharides in vitro. This capacity is reflected in its genome which has the largest repertoire of genes involved in acquisition and metabolism of polysaccharides among gut microbes, these include paralogous groups that are involved in polysaccharide uptake and degradation (glycosylhydrolases, cell-surface carbohydrate-binding proteins) (Sonnenburg et al. 2005). The representation of predicted glycosylhydrolases (α-galactosidases, β-galactosidases, α-glucosidases, β -glucosidases, β -glucuronidases, βfructofuranosidases, α -mannosidases, amylases, and endo-1,2-β-xylanases, plus 14 other activities. It has been determined that approximately 61% of its glycosylhydrolases are located in the periplasmic space or in the external membrane, which allows us to infer that these proteins contribute significantly to the metabolic medium of the intestinal environment, conditioning other species present in the same way (Xu., J. et al., 2003). *B. thetaiotamicron,* has also evolved the capacity to use a variety of host-derived glycans, including chondroitin sulfate, mucin, hyaluronate and heparin (Xu J. et al., 2003; Sonnenburg JL., et al., 2005), where a gnotobiotic mouse was used as a study model to simplify the conditions of the intestinal ecosystem present in humans, in such a way that it was determined that *B. thetaiotaomicron* modifies its carbohydrate uptake mechanisms from the polysaccharides of the diet of the hosts based on nutrient availability. Another interesting species in this sense, are those belonging to the genus *Akkermansia. A. municiphila* growth has been observed in the presence of N-acetylgalactosamine. In this sense, the absorption of these carbohydrates by one part and glucose in the absence of mucin, improved significantly with the addition of mucin, an element that indicates that additional elements derived from the metabolism of mucin are required as cofactors for optimal growth. On the other hand, in other experiments *A. municiphila* was cultivated under conditions in the presence of mucin, expression profiles were analyzed through the study of the transcriptome (RNA-Seq), where it was possible to confirm the activity of genes related to the degradation of mucin, and where the induction of these genes by the presence of mucin could be determined. In this same sense, it was possible to demonstrate through the subsequent analysis of the proteome the preference of this bacterium for the degradation of this type of carbohydrates, from which the relevance of these sugars in the phenomena of adaptation to the intestinal mucosa could be inferred. These studies confirm prevalent analyzes in the sense of considering *A. municiphila* a specialist organism for mucin in the intestine (Ottman, et al., 2017). In other relevant studies, metatranscriptomic analysis was used to delve into the analysis of bacterial interaction. Trophic-type interactions between species of special relevance in the mucosa of *A. muciniphila* and *A. caccae* were studied, where it was determined that butyrate associated with the host epithelium was produced as a result of this interaction. On the other hand, changes in the intensity of the *A. muciniphila* response were observed as a consequence of the presence of *A. caccae*, and where the role of *A. muciniphila* as the main species involved in the regulation of mucolytic activity for the maintenance of intestinal environmental conditions (Loo Wee Chia et al., 2018).

5. COEXISTENCE AND INTERACTION BETWEEN MICROORGANISMS IN THE INTESTINE AS A SPECIFIC ADAPTATION MECHANISM

The microorganisms that are present in the intestinal tract of mammals represent hundreds of different species, the interactions that take place within communities and with the host are very important for their adaptation and successful colonization. (Bivar X.K., 2018) The Quorum sensing (QS) that is established among the diversity of microbial components regulates behavior, generating chemical signaling molecules, producing a collective behavior and gene expression in these microbial communities. (Papenfort & Bassler 2016) The normal QS mechanism can be grouped into: production of the biochemical signal, accumulation of the signal, detection of the signal and degradation or inactivation of the QS signal. We can find three main categories of autoinductors on which the QS is based, those based on N-Acyl homoserine lactons (AHL) which is the most common in gram negative bacteria, the substrate S-adenosymethionine is used as a precursor in the biosynthesis of AHL´s together with acylated acyl carrier protein. Also, those that are based on Peptides that are mainly used by Gram-negative bacteria and auto-inducers-2 (AI-2) used by both Gram-positive and Gram-negative bacteria. (LaSarre & Federle 2013). There are a wide variety of phenotypes associated with QS such as: the formation of biofilms, as an adaptation mechanism in the face of constant changes in the intestinal tract with receptors (RhIR) and synthases (SwrR), motility, virulence and colonization of the host. (LaSarre & Federle 2013). Darkoh et al., (2019) report that the persistence of *Clostridium difficile* causing microbial dysbiosis in the human intestinal tract, may be associated with the production of indole, a bioactive molecule that can inhibit the growth of the protective microbiota. This indole is produced from tryptophan by the hydrolytic enzyme tryptophanase, producing Indole, pyruvate, and ammonia. This molecule of indole provides C. difficile with the necessary conditions to proliferate and persist. Much of the success in the adaptation and colonization processes of the microorganisms present in the intestinal tract is based on the chemical communication provided by the QS, and it is increasingly frequent that through the use of the -omics, more studies can be accessed to understand the QS.

CONCLUSION

In general, it is possible to say that the molecular mechanisms and bases that explain the adequacy phenomena in the intestinal environment must be studied considering the intrinsic complexity of the intestinal environment. In this sense, it is necessary to emphasize the large number of microorganisms present in the microbiota, a fact that from the quantitative point of view is very relevant because it involves a large number of interactions, which, in the first place, constitutes a reservoir of genetic diversity, since under these conditions, the mechanisms of horizontal gene transfer (mainly conjugation and transduction) are maximized by two factors, the diversity of microenvironments in the intestine, which generates an enormous variety of ecological niches and two by diet variability. On the other hand, another of the elements that should be highlighted as a factor that contributes to optimizing the phenomena of bacterial adaptation are the feedback effects that occur between relevant species of the microbiota and the host. In this sense, it is pertinent to allude to the interactions of a mutualistic type that are very numerous and where benefits in the hosts in terms of nutrition and immunology have been published. Finally, it is pertinent to include a clear perspective regarding the focus of the studies in this line of research, which can be summarized in the need to use -omics techniques to pose new research questions, since these methodologies implicitly focus on the analysis of complex systems and environments such as the microbiota must be studied under this paradigm if what is intended is to answer or generate new research questions regarding the macroscopic effects that occur in the physiology of the host organisms.

REFERENCES

Albenberg, L., Esipova, T. V., Judge, C. P., Bittinger, K., Chen, J., Laughlin, A., Grunberg, S., Baldassano, R. N., Lewis, J. D., Li, H., Thom, S. R., Bushman, F. D., Vinogradov, S. A., Wu, G. D. (2014). Correlation between intraluminal oxygen gradient and radial partitioning of intestinal microbiota. *Gastroenterology, 147(5),* 1055–63.e8. https://doi.org/10.1053/j.gastro.2014.07.020.

Bivar, X.K. (2018). Bacterial interspecies quorum sensing in the mammalian gut microbiota [published correction appears in *Comptes Rendus Biologies, 341(5),* 297-299]. https://doi.org/10.1016/j.crvi.2018.03.006.

Bhalodi, A. A., van Engelen, T., Virk, H. S., & Wiersinga, W. J. (2019). Impact of antimicrobial therapy on the gut microbiome. *The Journal of antimicrobial chemotherapy*, 74(Suppl 1), i6–i15. https://doi.org/10.1093/jac/dky530.

Cani, P.D. (2018). Human gut microbiome: hopes, threats and promises. *Gut 67(9),* 1716-1725. https://doi.org/10.1136/gutjnl-2018-316723.

Chia L.W., Hornung B.V., Aalvink S., Schaap P.J., de Vos W.M., Knol J., Belzer, C. (2018). Deciphering the trophic interaction between Akkermansia muciniphila and the butyrogenic gut commensal Anaerostipes caccaeusing a metatranscriptomic approach. *Antonie van*

Leeuwenhoek 111, 859–873. https://doi.org/10.1007/s10482-018-1040-x.

Corfield, A. Shukla, A.K. (2003). Mucins: Vital components of the mucosal defensive barrier. *Genomic/Proteomic Technol*. 3. 20-22.

Corfield, A. P. (2018). The Interaction of the Gut Microbiota with the Mucus Barrier in Health and Disease in Human. *Microorganisms 6(3)*, 78. https://doi.org/10.3390/microorganisms6030078.

David, L. A., Maurice, C. F., Carmody, R. N., Gootenberg, D. B., Button, J. E., Wolfe, B. E., Turnbaugh, P. J. (2014). Diet rapidly and reproducibly alters the human gut microbiome. *Nature 505,* 559-563. https://doi.org/10.1038/nature12820.

Darkoh, C., Plants-Paris K., Bishoff, D., DuPont H.L. (2019). Clostridium difficile Modulates the Gut Microbiota by Inducing the Production of Indole, an Interkingdom Signaling and Antimicrobial Molecule *mSystems 4(2),* e00346-18; https://doi.org/10.1128/mSystems.00346-18.

Everard, A., Belzer, C., Geurts, L., Ouwerkerk, J.P., Druart, C., Bindels, L.B., Guiot, Y., Derrien, M., Muccioli, G.G., Delzenne, N.M., de Vos, W.M., Cani. P.D. (2013). Cross-talk between Akkermansia muciniphila and intestinal epithelium controls diet-induced obesity. *Proceedings of the National Academy of Sciences of the United States of America, 110(22),* 9066-9071. https://doi.org/10.1073/pnas.1219451110.

Freter, R., Brickner, H., Botney, M., Aranki, A. (1983). Mechanisms that control bacterial populations in continuous-flow culture models of mouse large intestinal flora. *Infect. Immun. 39(2)*, 676–85.

Garud, N. R., Good, B. H., Hallatschek, O., & Pollard, K. S. (2019). Evolutionary dynamics of bacteria in the gut microbiome within and across hosts. *PLoS biology, 17(1),* e3000102. https://doi.org/ 10.1371/journal.pbio.3000102.

Harris, V. C., Armah, G., Fuentes, S., Korpela, K. E., Parashar, U., Victor, J. C., Tate, J., de Weerth, C., Giaquinto, C., Wiersinga, W. J., Lewis, K. D., & de Vos, W. M. (2017). Significant Correlation Between the Infant Gut Microbiome and Rotavirus Vaccine Response in Rural

Ghana. *The Journal of infectious diseases*, *215*(1), 34–41. https://doi.org/10.1093/infdis/jiw518.

Kelly, D., Mulder, I.E. (2012). Microbiome and immunological interactions *Nutrition Reviews, 70 (Suppl. 1),* S18-S30. https://doi.org/10.1111/j.1753-4887.2012.00498.x.

LaSarre, B., Federle, M.J. (2013). Exploiting Quorum Sensing To Confuse Bacterial Pathogens *Microbiology and Molecular Biology Reviews, 77 (1)* 73-111. https://doi.org/10.1128/MMBR.00046-12.

Lerner, A., Matthias, T., Aminov, R. (2017). Potential Effects of Horizontal Gene Exchange in the Human Gut. *Frontiers in immunology, 8,* 1630. https://doi.org/10.3389/fimmu.2017.01630.

Li, H., Limenitakis, J., Fuhrer, T., Geuking M.B., Lawson M.A., Wyss M., Brugiroux S., Keller I., Macpherson J.A., Rupp S., Stolp B., Stein J.V, Stecher B., Sauer, U., McCoy K.D., Macpherson A.J. (2015) The outer mucus layer hosts a distinct intestinal microbial niche. *Nature Communications 6*, 8292. https://doi.org/10.1038/ncomms9292.

Lillehoj, E. P., Kato, K., Lu, W., Kim, K. C. (2013). Cellular and molecular biology of airway mucins. *International review of cell and molecular biology, 303,* 139–202. https://doi.org/10.1016/B978-0-12-407697-6.00004-0.

Ottman, N., Davids, M., Suarez-Diez, M., Boeren, S., Shaap, P.J., Martin dos Santos, A.P., Smidt, H., Belzer, C., de Vos, W.M. (2017). Genome-Scale Model and Omics Analysis of Metabolic Capacities of Akkermansia muciniphila Reveal a Preferential Mucin-Degrading Lifestyle. *Applied and environmental microbiology, 83(18),* e01014-17. https://doi.org/10.1128/AEM.01014-17.

Papenfort, K., Bassler, B. (2016). Quorum sensing signal–response systems in Gram-negative bacteria. *Nat. Rev. Microbiol. 14*, 576–588. https://doi.org/10.1038/nrmicro.2016.89.

Pereira, F.C., Berry D. (2017). Microbial nutrient niches in the gut. *Environ. Microbiol.* 19(4). 1366-1378. https://doi.org 10.1111/1462-2920.13659.

Pickard, J. M., Zeng, M. Y., Caruso, R., Núñez, G. (2017). Gut microbiota: Role in pathogen colonization, immune responses, and inflammatory disease. *Immunological reviews, 279(1),* 70–89. https://doi.org/10.1111/imr.12567.

Plichta, D.R., Juncker, A.S., Bertalan, M., Rettedal, E., Gautier, L., Varela, E., (2016). Transcriptional interactions suggest niche segregation among microorganisms in the human gut. *Nature Microbiology* 1(11): 16152. https://doi.org doi: 10.1038/nmicrobiol.2016.152.

Qin, J., Li, R., Raes, J., Arumugam, M., Burgdorf, K. S., Manichanh, C., Nielsen, T., Pons, N., Levenez, F., Yamada, T., Mende, D. R., Li, J., Xu, J., Li, S., Li, D., Cao, J., Wang, B., Liang, H., Zheng, H., Xie, Y., Wang, J. (2010). A human gut microbial gene catalogue established by metagenomic sequencing. *Nature, 464(7285),* 59–65. https://doi.org/10.1038/nature08821.

Rinninella, E., Raoul, P., Cintoni, M., Franceschi, F., Miggiano, G., Gasbarrini, A., & Mele, M. C. (2019). What is the Healthy Gut Microbiota Composition? A Changing Ecosystem across Age, Environment, Diet, and Diseases. *Microorganisms, 7(1),* 14. https://doi.org/10.3390/microorganisms7010014.

Savignac, H. M., Corona, G., Mills, H., Chen, L., Spencer, J. P., Tzortzis, G. (2013). Prebiotic feeding elevates central brain derived neurotrophic factor, N-methyl-D-aspartate receptor subunits and D-serine. *Neurochem. Int. 63,* 756–764. https://doi.org/10.1016/j.neuint.2013.10.006.

Shoaie, S., Ghaffari, P., Kovatcheva-Datchary, P., Mardinoglu, A., Sen, P., Pujos-Guillot, E., de Wouters, T., Juste, C., Rizkalla, S., Chilloux, J., Hoyles, L., Nicholson, J. K., MICRO-Obes Consortium, Dore, J., Dumas, M. E., Clement, K., Bäckhed, F., Nielsen, J. (2015). Quantifying Diet-Induced Metabolic Changes of the Human Gut Microbiome. *Cell metabolism, 22(2),* 320–331. https://doi.org/10.1016/j.cmet.2015.07.001.

Sonnenburg, J. L., Xu, J., Leip, D. D., Chen, C. H., Westover, B. P., Weatherford, J., Buhler, J. D., Gordon, J. I. (2005). Glycan foraging in vivo by an intestine-adapted bacterial symbiont. *Science (New York, N.Y.), 307*(5717), 1955–1959. https://doi.org/10.1126/science.1109051.

Tamura, K., Hemsworth, G. R., Déjean, G., Rogers, T. E., Pudlo, N. A., Urs, K., Jain, N., Davies, G. J., Martens, E. C., Brumer, H. (2017). Molecular Mechanism by which Prominent Human Gut Bacteroidetes Utilize Mixed-Linkage Beta-Glucans, Major Health-Promoting Cereal Polysaccharides. *Cell reports, 21(2)*, 417–430. https://doi.org/10.1016/j.celrep.2017.09.049.

Thursby, E., Juge, N. (2017). Introduction to the human gut microbiota. *The Biochemical journal, 474(11)*, 1823–1836. https://doi.org/10.1042/BCJ20160510.

Weston R. Whitaker, Elizabeth Stanley Shepherd, Justin L. (2017). Sonnenburg, Tunable Expression Tools Enable Single-Cell Strain Distinction in the Gut Microbiome. *Cell, 169(3)*, 538-546.e12, https://doi.org/10.1016/j.cell.2017.03.041.

Xu, J., Bjursell, M. K., Himrod, J., Deng, S., Carmichael, L. K., Chiang, H. C., Hooper, L. V., Gordon, J. I. (2003). A genomic view of the human-Bacteroides thetaiotaomicron symbiosis. *Science, 299*(5615), 2074–2076. https://doi.org/10.1126/science.1080029.

Yamada, T., Hino, S., Iijima, H., Genda, T., Aoki, R., Nagata R., Han, K-H., Hirota, M., Kinashi Y., Oguchi, H., Suda, W., Furusawa, Y., Fujimura, Y., Kunisawa J., Hattori, M., Fukushima, M., Morita, T. Hase, K. (2019). Mucin O-glycans facilitate symbiosynthesis to maintain gut immune homeostasis. *EBioMedicine, 48*, 513-525. https://doi.org/10.1016/j.ebiom.2019.09.008.

Yang, I., Corwin, E. J., Brennan, P. A., Jordan, S., Murphy, J. R., Dunlop, A. (2016). The Infant Microbiome: Implications for Infant Health and Neurocognitive Development. *Nursing research, 65(1)*, 76–88. https://doi.org/10.1097/NNR.0000000000000133.

In: Molecular Basis of Specific Mechanism ... ISBN: 978-1-53618-751-9
Editors: Marcos López-Pérez et al. © 2020 Nova Science Publishers, Inc.

Chapter 7

MOLECULAR BASIS OF ADAPTATION AND MECHANISMS USED BY HALOPHILIC BACTERIA

Luis Mario Hernández Soto
and José Félix Aguirre Garrido[*]
Environmental Sciences Department, Metropolitan Autonomous
University (Lerma Unit) Lerma de Villada, México

ABSTRACT

The extreme ecosystems are distributed in different places in the biosphere, of particular relevance are the halophyte environments, since they are widely distributed, from the oceans to the brackish systems occupying the largest surface on the planet. Considering this, it is pertinent to allude to the enormous biological diversity adequate to these conditions that are the substrate for eventual technological developments. This chapter delves into the molecular bases of the most relevant bacterial mechanisms that allow these microorganisms to survive in these ecosystems.

Keywords: adaptation mechanism, halophilic, bacteria

---

[*] Corresponding Author: José Félix Aguirre Garrido PhD, Environmental Sciences Department, Metropolitan Autonomous University (Lerma Unit) Lerma de Villada, México. Email: j.aguirre@correo.ler.uam.mx; Tel.: (+52-728) 2822785 Ext. 1011.

1. INTRODUCTION

Ecosystems classified as extreme from a human point of view (those which present "abnormal" ranges in temperature, pH, or salt concentrations) typically show a lack of "higher" forms of life; despite of this, they are inhabited by a relatively abundant microbial community, that has developed adaptations to these ecological niches (Trüper & Galinski, 1986). One extreme condition to survive in some ecosystems, is high salt concentrations. Main problem in these systems is that water activity (water that is available for organic intake) is inversely proportional to the salt concentration (Galinski, 1995a; Shivanand & Mugeraya, 2011). These osmotic challenging conditions are not reserved to ionic environments, it is possible to find them in non-ionic solutions like syrups, honey, etc. Environments like the ones we referred before are known as low water ones. Due to cellular membranes are semi permeable to water molecules, differences between ions concentrations outside and inside the cell, are rapidly favored to the one with the higher concentration; this will lead to subsequent dehydration and it will stop its growth until the organism get some adaptation to its environment. A halophile organism is defined as the one that needs high salt concentrations for its growing. The most accepted and spread definition of these organisms was proposed for Kushner and Kamekura (Kushner, 1978; Kushner & Kamekura, 1988), they cataloged them according to their necessity of salt concentration for its optimum growth. Therefore, non-halophiles grow easily in media with less than 0.2 M of any salt, on the other hand, halophiles grow in media with ranges of salts that go from 0.2 to 5.2 M. In addition to it, halophiles organisms can be classified in: (i) slightly halophiles (most rapid growth at 2 to 5% NaCl or 0.34 to 0.85 M), (ii) moderately halophiles (most rapid growth at 5 to 20% NaCl or 0.85 to 3.4 M), and (iii) extremely halophiles (most rapid growth at 20 to 30% NaCl or 3.4 to 5.1 M) (Antón, 2011; Kushner, 1978; Ollivier et al., 1994; Ventosa et al., 1998). We can find halophile organisms in all the life domains, among *Bacteria* we can find representative members from Proteobacteria, Firmicutes, Actinobacteria, Bacteroidetes and Cyanobacteria *phyla*, meanwhile the most important in

Archaea are microorganisms from the Halobacteria class (Oren, 2008). In *Eukarya* domain there are different organisms that includes fungus in orders as Wallemiales, Trichnosporales, Capnodiales and Sporidiales (Arakaki et al., 2013; Plemenitaš et al., 2014); algae like Dunaliellaceae species and a few animals as *Artemia salina* (Faraj Edbeib et al., 2016; Harding et al., 2016a). Salt requirement and tolerance are parameters erratic among species. Even more, these parameters can be variable, due to they could change directly influenced for the growth temperature and the nature of the nutrients available (Kushner, 1993). Halophiles are different from halotolerant in their dependency of NaCl for growing; meanwhile, second ones are capable to deal with fluctuations of saline concentrations at their environments, and salt is not an obligated condition for growing. In contrast to halotolerant bacteria, which do not require NaCl for growth but can grow under saline conditions, halophiles must have NaCl for growth (Ollivier et al., 1994).

2. STRATEGIES

In order to survive to high salinity of their environment, bacteria, need to equalize their cytoplasm with the external medium. It means that they must find a way to deal with changing osmotic pressures (Imhoff, 1993). This balance can be reached using different strategies from the inherent composition to their membrane composition, e.g., it has been reported that gram positive bacteria have better tolerance to osmotic stress even without specialized strategies against it (Galinski, 1995a). Some studies suggest that some organisms (e.g., *Salinivibrio costicola*) have the ability to control intake and outtake of water in order to get a hypoosmotic environment at their intracellular space (Shindler et al., 1977; Ventosa et al., 1998; Vreeland, 1987); but mainly, there are two main approaches adapted by the microorganisms to cope with the osmotic stress of high salt concentrations accumulating different salts, organic molecules or often a combination of them.

3. SALTING IN

If an organism develop the strategy of maintain high intracellular salt quantities conditions, in order to get a minimum osmotic shock compared to the external concentrations it is using the strategy knows as "salting-in" (Galinski, 1995b; A Oren, 2002; Oren et al., 2013). In this way to deal with stress, all intracellular systems should be adapted to high salt concentrations. It is relevant to point, that potassium ions are preferred for this, rather than sodium ones (Gunde-Cimerman et al., 2018; Oren, 2002). Some transporters as TrkH and TrkI are responsible of potassium intake in *Halomonas elongate* (Kraegeloh et al., 2005). From an energetic point of view, this approach is relative cheap spending just two molecules of ATP for every three potassium ions, but it implicates way more proteins' adaptations in order to be active and work properly at those levels of salt, in other words, for this to happen all intracellular systems should be adapted to high salt concentrations. To achieve this, it has been demonstrated that those organisms that "salt-in" owns proteins rich in acidic and hydrophilic residues. It leads to very acidic proteomes, with high presence of aspartate and glutamate and a very low one of lysine at their protein surface (Harding et al., 2016a; Oren, 1999; Paul et al., 2008). As a consequence of all mentioned, high salt levels results in a raise of hydrophobicity, this probably gives halophilic proteins the property of avoiding rigid folded conformation and aggregation; making them quite more stables than other proteins that surely would aggregate in this kind of environments; and also preserve a hydration shell resulting in solubility of proteins (Harding et al., 2016a; Mevarech et al., 2000; Weinisch et al., 2018). This strategy transforms proteins so deeply that they shows a significant instability in environments with low salt concentrations (Siglioccolo et al., 2011). Although, salting-in strategists are reported as the most extreme halophiles in nature (Oren, 2008; Oren et al., 2013), they are greatly dependent to high levels of salt and require wide adaptations of the whole intracellular machinery, including enzymes, structural proteins, and charged amino acids that allow the retention of water molecules on their surfaces (Raval et al., 2018). This "transformation" is only reachable after a long and complex evolutionary process (Dennis & Shimmin, 1997). Thus, hypoosmotic conditions are disastrous for this hyperspecialized organisms, provoking systematic destabilization and therefore, malfunctions of proteins, then, only a few groups of organisms have chosen this survival strategy, such as Halobacteriaceae (*Archaea*), *Salinibacter* (*Bacteria*), and Halanaerobiales (*Bacteria*) (Gunde-Cimerman et al., 2018; Weinisch et al., 2018).

4. SALTING OUT, OR COMPATIBLE SOLUTES STRATEGY

More spread strategy, used by some halophilic and all halotolerant organisms is the opposite to salting in, that means they spend energy eliminating salt from their cytoplasm, in order to avoid aggregation of their proteins. This is "salting out" or better known as "compatible solute strategy" (Brown, 1990; Gohel et al., 2015; Weinisch et al., 2018; Welsh, 2000). Here, the osmotic pressure is equalized by the synthesis (or intake) of organic compatible solutes. This allows the lack of further adaptations of the intracellular systems (Gohel et al., 2015). Salt-out strategists need to maintain a stable high intracellular K^+/Na^+ ratio for several physiological functions like osmotic regulation, protein synthesis, enzyme activation or the maintenance of the plasma membrane potential (Rodríguez-Navarro, 2000; Weinisch et al., 2018) Osmoregulation in salting out uses way less potassium ions than salting in, and it is carrying out mainly by mixing them with organic solutes (compatible solutes, osmolites or osmoprotectants). These are uncharged or zwitterionic low molecular mass organic (Figure 1) compounds that provides osmotic balance, maintain cell turgor, and protect non-salt adapted proteins; they do not carry a net charge at physiological pH (Brown, 1990; Oren, 1999; Roberts, 2005; Weinisch et al., 2018; Welsh, 2000). In addition to this, they do not interfere with the cell metabolism at high cytoplasmatic concentrations; they are compatible with cellular metabolism (hence the name), so they do not disturb important cell processes (DNA replication, interaction DNA-protein, etc.)

as it could happen with inorganic ions (Kempf & Bremer, 1998; Record et al., 1998).

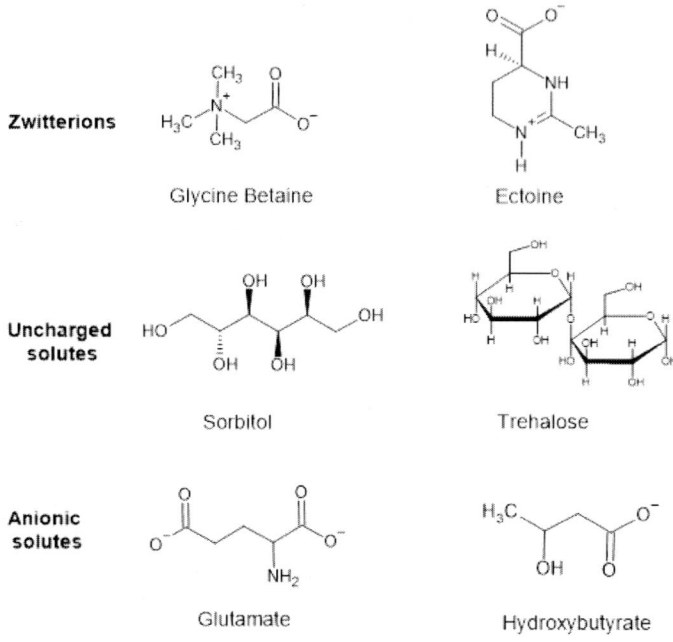

Figure 1. Most common compatible solutes synthetized by halotolerant and halophilic organisms.

Some compatible solutes are synthetized *de novo* within the cell, others can be taken from their environments. Common compatible solutes found in halophilic organisms include amino acids and polyols (glycine, betaine, ectoine, sucrose, trehalose, and glycerol) (Raval et al., 2018). Some of these molecules, especially glycine, betaine, and ectoines, are important in different biotechnology fields (Ventosa et al., 1998). Glycine betaine, proline, and trehalose are widely used to deal with moderately to high osmotic stress that shows around 1 M of NaCl (Raval et al., 2018). In addition to its change in osmotic pressure, compatible solutes are able to prevent other damages (unfolding of proteins) in cellular structures caused by heating, freezing or drying conditions (Galinski, 1995b).

Salting out strategists show a higher tendency to adaptation in a wide range of salt concentrations, specially if we compare them against salting in strategists (Gohel et al., 2015; Ventosa et al., 1998). Energetically, this ability is a pretty expensive one. To synthesize a single molecule a generic compatible solute, an autotrophic cell spends a range of 30 – 90 ATP molecules, meanwhile a heterotrophic one needs 23 – 79 ATP molecules (Raval et al., 2018). Moreover, this strategy is highly dependent to salinity intensity, osmotic stress duration, surrounding osmolytes or available carbon source and raw material availability (Roberts, 2005). It has been observed a clear tendency in the kind of compatible solutes produced by slightly and moderately halophiles and halotolerants. First ones synthetize disaccharides, meanwhile second ones tend to produce and gather monosaccharides, polyols and nitrogen molecules (Kunte, 2006). Non-halophilic organisms as *Salmonella* collect trehalose or sucrose, thus its capacity for osmotic adaptation is limited to salt concentrations smaller than 0.5 M (Kempf & Bremer, 1998). Halotolerants have a smaller ability to deal with high salt concentration, as marine bacteria, produce sugar-polyols (glucosylglycerol). On the other hand, halophilic bacteria that can live with salt concentrations 1.72 M or more, tend to gather amino acids or derivatives (proline, betaine or ectoine). Production of compatible solutes is influenced for the abundance or scarcity of their precursors and adaptative plasticity inherent of each organism. (Kunte, 2006). Benefits from producing compatible solutes are more than just protect from salinity and dehydration. Sometimes they allow their producers to colonize different environments that could be hard to survive without them. It has been reported that production of glycine betaine in *Listeria monocytogenes* increases its cold tolerance (Ko et al., 1994); in the other hand, trehalose has been associated to thermotolerance in bacteria, yeast and fungi (Welsh, 2000). Even ectoine has been related to a significant increase in the rate of denitrification of some consortium (Cyplik et al., 2012). This shows how versatile compatible solutes can be, and that although its production is expensive, it is totally worth it. Previously we talk about acidity proteomes that salting in strategists have. Salting out organisms lack of that acidic proteins, this is curious because they synthetize extracellular proteins way more acidic and hydrophilic than the ones that mesophilic organisms produce (Harding et al., 2016b). Compatible solutes are quite water-soluble, this ensure a proper folding of proteins, therefore a regular performance in almost all cellular processes, keeping this in mind, no evolutionary changes in protein surfaces were needed at all (León et al., 2018).

CONCLUSION

As a conclusion that can be inferred after the analysis of the molecular bases of the adaptation processes to the presence of the salts in greater or lesser concentration, it refers to the fact that all of them converge in the maintenance of homeostatic regulation under intense pressure conditions osmotic, linking the processes of transport regulation at the membrane level with the phenomena of regulation of protein expression.

REFERENCES

Antón, J. (2011). Halophile. In M. Gargaud, R. Amils, J. C. Quintanilla, H. J. (Jim) Cleaves, W. M. Irvine, D. L. Pinti, & M. Viso (Eds.), *Encyclopedia of Astrobiology* (pp. 725–727). Springer Berlin Heidelberg. https://doi.org/10.1007/978-3-642-11274-4_694.

Arakaki, R. L., Monteiro, D. A., Boscolo, M., Dasilva, R., & Gomes, E. (2013). *Halotolerance, ligninase production and herbicide degradation ability of basidiomycetes strains.* www.sbmicrobiologia.org.br.

Brown, A. D. (1990). *Microbial water stress physiology. Principles and perspectives.* John Wiley & Sons.

Cyplik, P., Piotrowska-Cyplik, A., Marecik, R., Czarny, J., Drożdżyńska, A., & Chrzanowski, Ł. (2012). Biological denitrification of brine: the effect of compatible solutes on enzyme activities and fatty acid degradation. *Biodegradation*, *23*(5), 663–672.

Dennis, P. P., & Shimmin, L. C. (1997). Evolutionary divergence and salinity-mediated selection in halophilic archaea. *Microbiology and*

Molecular Biology Reviews : MMBR, 61(1), 90–104. https://pubmed. ncbi.nlm.nih.gov/9106366.

Faraj Edbeib, M., Roswanira, W. A., & Fahrul, H. (2016). Halophiles: biology, adaptation, and their role in decontamination of hypersaline environments. *World Journal of Microbiology and Biotechnology*, *32*(135). https://doi.org/10.1007/s11274-016-2081-9.

Galinski, E. A. (1995a). Osmoadaptation in Bacteria. In *Advances in Microbial Physiology* (Vol. 37, Número C). https://doi.org/10.1016/ S0065-2911(08)60148-4.

Galinski, E. A. (1995b). *Osmoadaptation in Bacteria* (R. K. B. T.-A. in M. P. Poole (ed.); Vol. 37, pp. 273–328). Academic Press. https://doi.org/ https://doi.org/10.1016/S0065-2911(08)60148-4.

Gohel, S. D., Sharma, A. K., Kruti G. Dangar, F., Thakrar, oram J., & Singh, S. P. (2015). Antimicrobial and biocatalytic potential of haloalkaliphilic actinobacteria. In D. K. Maheshwari & M. Saraf (Eds.), *Halophiles Biodiversity and Sustainable Exploitation Biodiversity and Sustainable Exploitation*. Springer, Cham. http://www.springer.com/series/11920.

Gunde-Cimerman, N., Plemenitaš, A., & Oren, A. (2018). Strategies of adaptation of microorganisms of the three domains of life to high salt concentrations. *FEMS Microbiology Reviews*, *009*, 353–375. https://doi.org/10.1093/femsre/fuy009.

Harding, T., Brown, M. W., Simpson, A. G. B., & Roger, A. J. (2016a). Osmoadaptative Strategy and Its Molecular Signature in Obligately Halophilic Heterotrophic Protists. *Genome Biol. Evol.*, *8*(7), 2241–2258. https://doi.org/10.1093/gbe/evw152.

Harding, T., Brown, M. W., Simpson, A. G. B., & Roger, A. J. (2016b). Osmoadaptative Strategy and Its Molecular Signature in Obligately Halophilic Heterotrophic Protists. *Genome Biology and Evolution*, *8*(7), 2241–2258. https://doi.org/10.1093/gbe/evw152.

Imhoff, J. F. (1993). Osmotic adaptation in halophilic and halotolerant microorganisms. In *The biology of halophilic bacteria* (pp. 87–103). CRC Press Inc.

Kempf, B., & Bremer, E. (1998). Uptake and synthesis of compatible solutes as microbial stress responses to high-osmolality environments. *Archives of Microbiology volume*, *170*, 319–330.

Ko, R., Smith, L. T., & Smith, G. M. (1994). Glycine Betaine Confers Enhanced Osmotolerance and Cryotolerance on *Listeria monocytogenes*. *Journal of Bacteriology*, *176*(2), 426–431.

Kraegeloh, A., Amendt, B., & Kunte, H. J. (2005). Potassium Transport in a Halophilic Member of the Bacteria Domain: Identification and Characterization of the K Uptake Systems TrkH and TrkI from *Halomonas elongata* DSM 2581 T. *Journal of Bacteriology*, *187*(3), 1036–1043. https://doi.org/10.1128/JB.187.3.1036-1043.2005.

Kunte, H. J. (2006). Osmoregulation in bacteria: Compatible solute accumulation and osmosensing. *Environmental Chemistry*, *3*(2), 94–99. https://doi.org/10.1071/EN06016.

Kushner, D. J. (1978). Life in high salt and solute concentrations: halophilic bacteria. In D. J. Kushner (Ed.), *Microbial Life in Extreme Environments* (pp. 317–368). Academic Press. http://ci.nii.ac.jp/naid/10016643410/en/.

Kushner, D. J. (1993). Growth and nutrition of halophilic bacteria. In R. H. Vreeland & L. I. Hochstein (Eds.), *The biology of halophilic bacteria* (pp. 87–103). CRC Press Inc.

Kushner, D. J., & Kamekura, M. (1988). Physiology of halophilic eubacteria. In F. Rodríguez-Valera (Ed.), *Halophilic Bacteria* (pp. 109–138). CRC Press. http://ci.nii.ac.jp/naid/10013365431/en/.

León, M. J., Hoffmann, T., Sánchez-Porro, C., Heider, J., Ventosa, A., & Bremer, E. (2018). Compatible solute synthesis and import by the moderate halophile *Spiribacter salinus*: Physiology and genomics. *Frontiers in Microbiology*, *9*(FEB), 1–18. https://doi.org/10.3389/fmicb.2018.00108.

Mevarech, M., Frolow, F., & Gloss, L. M. (2000). Halophilic enzymes: Proteins with a grain of salt. *Biophysical Chemistry*, *86*(2–3), 155–164. https://doi.org/10.1016/S0301-4622(00)00126-5.

Ollivier, B., Caumette, P., Garcia, J. L., & Mah, R. A. (1994). Anaerobic bacteria from hypersaline environments. *Microbiological Reviews*, *58*(1), 27–38. https://doi.org/10.1128/mmbr.58.1.27-38.1994.

Oren, A. (1999). Bioenergetic Aspects of Halophilism. In *Microbiology and molecular biology reviews* (Vol. 63, Número 2).

Oren, A. (2002). Diversity of halophilic microorganisms: Environments, phylogeny, physiology, and applications. *Journal of Industrial Microbiology & Biotechnology*, *28*, 56–63. https://doi.org/10.1038/sj/jim/7000176.

Oren, Aharon. (2008). Saline Systems. *Saline Systems*, *4*(2). https://doi.org/10.1186/1746-1448-4-2.

Oren, Aharon, Ventosa, A., Amoozegar, M. A., & Mormile, M. R. (2013). *Life at high salt concentrations, intracellular KCl concentrations, and acidic proteomes.* https://doi.org/10.3389/fmicb.2013.00315.

Paul, S., Bag, S. K., Das, S., Harvill, E. T., & Dutta, C. (2008). Molecular signature of hypersaline adaptation: insights from genome and proteome composition of halophilic prokaryotes Molecular signatures of halophilic prokaryotes A comparative genomic and proteomic study of halophilic and non-halophilic prokaryotes. *Genome Biology*, *9*(4), 70. https://doi.org/10.1186/gb-2008-9-4-r70.

Plemenitaš, A., Lenassi, M., Konte, T., Kejžar, A., Zajc, J., Gostinčar, C., Gunde-Cimerman, N., Oren, A., & Hebrew, T. (2014). *Adaptation to high salt concentrations in halotolerant/halophilic fungi: a molecular perspective.* https://doi.org/10.3389/fmicb.2014.00199.

Raval, V. H., Bhatt, H. B., & Satya P. Singh. (2018). Adaptation Strategies in Halophilic Bacteria. In R. V. Durvasula & D. V. Subba Rao (Eds.), *Extremophiles from biology to biotechnology* (pp. 138–150). CRC Press. https://doi.org/https://doi.org/10.1201/9781315154695.

Record, M. T., Courtenay, E. S., Cayley, D. S., & Guttman, H. J. (1998). Responses of E. coli to osmotic stress: Large changes in amounts of cytoplasmic solutes and water. *Trends in Biochemical Sciences*, *23*(4), 143–148. https://doi.org/10.1016/S0968-0004(98)01196-7.

Roberts, M. F. (2005). Organic compatible solutes of halotolerant and halophilic microorganisms. *Saline Systems*, *1*(5). https://doi.org/10.1186/1746-1448-1-5.

Rodríguez-Navarro, A. (2000). Potassium transport in fungi and plants. *Biochimica et Biophysica Acta - Reviews on Biomembranes*, *1469*(1), 1–30. https://doi.org/10.1016/S0304-4157(99)00013-1.

Shindler, D. B., Wydro, R. M., & Kushner, D. J. (1977). Cell-Bound Cations of the Moderately Halophilic Bacterium *Vibrio costicola*. *Journal of Bacteriology*, *130*(2), 698–703. http://jb.asm.org/.

Shivanand, P., & Mugeraya, G. (2011). Halophilic bacteria and their compatible solutes -osmoregulation and potential applications. *Current Science*, *100*(10), 1516–1521.

Siglioccolo, A., Paiardini, A., Piscitelli, M., & Pascarella, S. (2011). *Structural adaptation of extreme halophilic proteins through decrease of conserved hydrophobic contact surface*. https://doi.org/10.1186/1472-6807-11-50.

Trüper, H. G., & Galinski, E. A. (1986). Concentrated brines as habitats for microorganisms. *Experientia*, *42*(11), 1182–1187. https://doi.org/10.1007/BF01946388.

Ventosa, A., Nieto, J., & Oren, A. (1998). Biology of Moderately Halophilic Aerobic Bacteria. In *Microbiology and molecular biology reviews* (Vol. 62, Número 2). http://mmbr.asm.org/.

Vreeland, R. H. (1987). Mechanisms of Halotolerance in Microorganisms. *CRC Critical Reviews in Microbiology*, *14*(4), 311–356. https://doi.org/10.3109/10408418709104443.

Weinisch, L., Kü hner, S., Roth, R., Grimm, M., Roth, T., A Netz, D. J., Pierik, A. J., & Filker, S. (2018). Identification of osmoadaptive strategies in the halophile, heterotrophic ciliate *Schmidingerothrix salinarum*. *PLoS Biology*, *16*(1), 1–29. https://doi.org/10.1371/journal.pbio.2003892.

Welsh, D. T. (2000). Ecological significance of compatible solute accumulation by micro-organisms: from single cells to global climate. *FEMS Microbiology Reviews*, *24*(3), 263–290. https://doi.org/10.1111/j.1574-6976.2000.tb00542.x.

ABOUT THE EDITORS

Marcos Lopez Perez PhD
Titular professor and researcher
Environmental Department. Biological and health Division, Autonomous Metropolitan University (Lerma Unit) Mexico

Born in Spain in 1978, graduated in Biological Sciences with a specialty in plant physiopathology and phytotechnology in 2001 and a master's degree in Molecular Biology of fungi in 2004 by the University of Salamanca, PhD in Biotechnology at the Metropolitan Autonomous University (2010) in the Iztapalapa Unit (Mexico), member of The National System of Researchers (SNI) since 2011. Dedicated to several lines of research in microbial ecology, physiology of *methylotrophic* yeast growth on inert solid supports, metagenomics of microbial communities in contaminated environments and resistance to antibotics, particularly in bacterial adaptation mechanisms inherently nonspecific, mainly proteins involved in cellular efflux pumps as well as in regulatory elements of its expression.

José Félix Aguirre Garrido
Member of the area of Biotechnology and Environmental Microbiology
Metropolitan Autonomous University, Campus Lerma

Dr. Aguirre-Garrido is Professor at the University Autonomous Metropolitan Unit Lerma, in the department of environmental sciences, research leader of the area of biotechnology and environmental microbiology. The work of Dr. Aguirre-Garrido's is supported by a total of 12 publications, an h-index of 5 and 112 citations according to the Scopus database. He is the co-author of 2 book chapters, has presented about 15 works in international academic events, 20 works in national congresses. In the period from January 2014 to January 2015, he did a postdoctoral research at the Zaidín in Spain, in the research group of Structure, Dynamics and Function of Rhizobacteria Genomes, during this stay he was related to the bioinformatic analysis of the projects assigned to the research group.

INDEX

A

access, 10, 40, 41, 43
acid, 14, 15, 21, 30, 51, 64, 65, 96, 99, 105, 121, 138
acidic, 14, 85, 134, 137, 141
active site, 84, 95, 97, 98, 101, 112
adaptation(s), vii, viii, 1, 2, 3, 5, 7, 8, 12, 13, 16, 19, 23, 51, 61, 62, 63, 66, 73, 76, 80, 86, 115, 118, 119, 120, 122, 123, 124, 131, 132, 134, 135, 136, 138, 139, 141, 142
adaptation mechanism, 2, 3, 5, 13, 16, 23, 67, 73, 124, 131
aggregation, 6, 7, 66, 134, 135
amino, 14, 21, 28, 51, 69, 85, 86, 95, 134, 136, 137
amino acid(s), 14, 21, 28, 51, 69, 85, 86, 95, 134, 136, 137
antibiotic, 4, 7, 9, 10, 11, 12, 17, 18, 19, 40, 42, 43, 46, 49, 51, 52, 53, 55, 56, 57, 58, 59, 68, 76, 96, 116
antibiotic resistance, 10, 17, 18, 19, 76
atoms, 87, 90, 91, 93
ATP, 14, 19, 134, 137

B

Bacillus subtilis, 70, 89, 96
bacteria, vi, vii, 1, 2, 3, 5, 7, 8, 10, 12, 13, 15, 16, 17, 19, 20, 21, 22, 41, 48, 50, 53, 55, 66, 69, 70, 75, 78, 79, 81, 84, 93, 96, 97, 116, 117, 118, 119, 121, 123, 126, 127, 131, 132, 133, 135, 137, 139, 140, 141, 142
bacterial, v, vii, 1, 2, 3, 5, 7, 8, 9, 12, 13, 17, 18, 19, 21, 22, 23, 41, 43, 46, 47, 56, 57, 58, 61, 62, 64, 65, 68, 69, 70, 73, 74, 75, 76, 78, 79, 80, 81, 85, 96, 97, 99, 100, 103, 104, 105, 115, 116, 119, 123, 124, 125, 126, 127, 128, 131
bacterial homeostasis, 62
bacterium, 20, 54, 64, 71, 75, 122
bioactive compounds, 41, 43
biochemical mechanisms, 39, 40
biodegradation, 100, 102, 103, 112, 120
biological processes, 44, 54, 84
bioremediation, 86, 94, 103, 113
biosphere, 1, 2, 61, 73, 131
biosynthesis, 30, 39, 40, 41, 42, 43, 46, 50, 51, 52, 53, 54, 57, 59, 80, 103, 123

biotechnology, 109, 136, 141
biotic, 1, 2, 12
BldD, 47, 52, 53, 54, 56, 57, 59

C

calcium, 26, 30, 37, 42, 52
carbohydrate(s), 97, 119, 121
carbon, 29, 32, 40, 43, 72, 91, 96, 106, 137
catalysis, 9, 11, 85, 97, 98, 99, 100, 103, 108
c-di-GMP, 53, 54, 57, 59
cellular homeostasis, 13, 16, 68, 81
challenges, 17, 46, 59, 102
chemical(s), 3, 6, 13, 30, 31, 39, 40, 63, 94, 99, 103, 109, 121, 123
chemotherapy, 21, 22, 125
classes, 45, 52, 69
cloning, 19, 74, 77, 79
clusters, 42, 54, 55, 57, 107
cluster-situated regulator(s), 42
colonization, 118, 119, 123, 127
color, 86, 89, 91
communication, 8, 15, 40, 124
community/communities, 7, 8, 23, 94, 116, 118, 123
complexity, vii, viii, 34, 111, 124
composition, 6, 15, 26, 62, 63, 79, 80, 115, 117, 133, 141
compounds, 8, 9, 39, 40, 41, 43, 49, 68, 78, 84, 99, 101, 107, 115, 118, 135
conidia, v, 25, 26, 27, 29, 31, 32, 34, 36, 37
conservation, 65, 90, 91
convergence, 7, 15, 16
copper, 13, 16, 18, 66, 74, 80, 84, 87, 88, 89, 90, 91, 92, 93, 94, 97, 98, 100, 105, 107, 109, 111, 112
culture, 26, 34, 36, 126
cytoplasm, 10, 14, 62, 67, 72, 133, 135

D

defects, 27, 30, 32, 64
degradation, 23, 30, 49, 50, 54, 76, 86, 96, 100, 101, 103, 106, 120, 121, 123, 138
detection, 16, 72, 123
detoxification, 66, 68, 78
diet, 71, 115, 116, 118, 120, 121, 124, 126
distribution, 1, 57, 73, 95, 117
diversity, vii, 2, 57, 66, 74, 79, 112, 118, 123, 124, 131
DNA, vii, 15, 33, 46, 48, 50, 51, 58, 86, 135
docking, 84, 86, 95, 98, 100, 103, 104, 106, 107, 108, 109, 112
drugs, 9, 65, 68, 86, 99, 111
dyes, 85, 96, 103, 112

E

E. coli, 14, 63, 68, 70, 141
ecosystem, viii, 5, 16, 23, 73, 117, 120, 122
effluents, 86, 96, 101
electron(s), 94, 97, 100, 101, 109
encoding, 10, 33, 34, 42
energy, 2, 6, 10, 13, 16, 23, 50, 51, 72, 99, 101, 107, 112, 116, 117, 120, 135
engineering, 50, 55, 85, 94, 98, 102, 110
entomopathogenic fungi, 25, 26, 27, 30, 34
environment(s), vii, 6, 7, 12, 14, 20, 35, 40, 49, 62, 77, 98, 101, 112, 118, 119, 122, 124, 125, 131, 132, 133, 134, 136, 137, 139, 140, 141
environmental change, 2, 4, 15, 39, 40, 63
environmental conditions, 5, 7, 8, 40, 48, 71, 72, 73, 118, 123
environmental stress, 2, 15, 31, 44
enzyme(s), viii, 9, 11, 13, 14, 16, 23, 26, 27, 30, 32, 33, 42, 46, 49, 54, 67, 75, 83, 84, 86, 89, 90, 94, 99, 101, 102, 105, 110, 111, 112, 124, 134, 135, 138, 140
epithelium, 121, 123, 126

evidence, 30, 50, 68
evolution, 4, 6, 15, 17, 18, 22, 59, 76, 86, 89, 90, 93, 96, 97, 99, 102, 107, 120
exposure, 6, 26, 40, 43, 66, 74

F

families, 28, 47, 74
fitness, 3, 18, 19
food, 20, 41, 85, 94, 99, 110, 115, 121
food industry, 41, 85, 110
force, viii, 93, 100
formation, 6, 31, 32, 44, 49, 54, 66, 77, 84, 89, 105, 123
fungus/fungi, 5, 22, 25, 26, 27, 28, 29, 30, 32, 31, 33, 34, 35, 36, 38, 84, 88, 93, 96, 97, 102, 110, 133, 137, 141, 142

G

gene expression, 28, 46, 50, 117, 123
gene transfer, vii, 67, 119, 124
genes, vii, 9, 10, 18, 19, 20, 26, 27, 30, 31, 32, 33, 34, 39, 40, 41, 42, 43, 44, 46, 47, 48, 49, 50, 51, 52, 53, 55, 63, 65, 68, 69, 79, 86, 116, 119, 120, 121
genome, 41, 42, 43, 44, 45, 51, 55, 56, 57, 89, 121, 141
genus, 23, 41, 43, 44, 46, 121
germination, 29, 30, 38
glucose, 51, 85, 122
glycans, vi, 115, 120, 122, 129
glycerol, 33, 38, 136
glycosylation, 12, 94, 97, 100, 106
gram-positive actinobacterium, 41
growth, 7, 27, 29, 30, 31, 32, 34, 37, 38, 42, 43, 44, 46, 47, 52, 56, 66, 80, 121, 122, 124, 132

H

halophilic, vi, 131, 134, 135, 136, 137, 138, 139, 140, 141, 142
health, 17, 116, 117
heavy metals, 2, 4, 12, 17, 66
histidine, 33, 38, 91, 93, 112
history, 1, 2, 5, 9, 26, 61, 65
homeostasis, 2, 13, 14, 16, 17, 18, 31, 49, 50, 62, 67, 68, 69, 76, 78, 81, 129
host, 15, 19, 29, 115, 117, 118, 120, 122, 123, 124
host diet, 115, 117
HrdB, 50, 51, 58
human, 40, 75, 99, 116, 118, 120, 124, 126, 128, 129, 132
hydrogen, 49, 99, 101, 108

I

identification, 9, 44, 66
induction, 31, 34, 49, 122
industry, viii, 44, 85, 96, 110, 111
infection, 16, 26, 117
inhibition, 7, 76, 100, 107, 109, 117
insects, 26, 84, 97
in-silico, 84, 87, 96, 97, 102, 104
integrity, 38, 49, 71
interface, 87, 89, 95
intestinal tract, 70, 117, 123
intestine, 116, 118, 119, 122, 124, 128
ions, 28, 69, 79, 98, 132, 134, 135
isozymes, 98, 100, 108

L

landscape, 3, 17, 56
landscapes, 3, 17, 18, 19
larvae, 26, 34, 36
lead, viii, 101, 132

life cycle, 26, 41, 44, 45
ligand, 9, 28, 30, 93, 96, 98, 100
light, 26, 28, 31, 32, 34, 35, 37, 87, 89
lignin, 85, 100, 103, 107
lipids, 33, 63, 64, 73, 75, 76, 80

motif, 22, 49, 87
mucus, 117, 121, 127
mutagenesis, 85, 86, 94, 99
mutation(s), 52, 94, 96, 97, 99, 119
mycelium, 26, 36, 44

M

magnesium, 12, 19, 69, 77
medicine, 21, 41, 80
membrane(s), v, 4, 10, 12, 15, 21, 28, 49, 61, 62, 63, 64, 65, 68, 69, 70, 71, 73, 74, 75, 76, 79, 80, 81, 121, 122, 132, 133, 135, 138
mercury, 67, 68, 78
messengers, 8, 16, 41, 43, 54, 57
metabolic pathways, 41, 58, 64
metabolism, 6, 13, 30, 33, 44, 46, 47, 50, 51, 58, 68, 77, 116, 121, 128, 135
metabolites, 2, 8, 30, 33, 39, 40, 44, 55, 57, 66, 68, 72
metals, 13, 67, 79
Mg^{2+}, 18, 69, 75, 76, 78, 80
microbial adaptation, 67, 115
microbial community/communities, 7, 116, 117, 123, 132
microbial natural products, 40
microbiota, 115, 116, 117, 118, 120, 124, 125, 127, 129
microorganism(s), vii, viii, 5, 8, 9, 14, 16, 25, 40, 44, 52, 57, 68, 69, 70, 115, 116, 117, 120, 123, 124, 128, 131, 133, 139, 141, 142
models, 94, 113, 126
modifications, 10, 64, 71, 85
molecular biology, 76, 127, 141, 142
molecular dynamics, 84, 86, 96, 100, 106, 108, 110
molecules, 10, 11, 30, 33, 41, 43, 53, 54, 70, 95, 98, 101, 123, 132, 133, 134, 136, 137
monomers, 89, 91, 121

N

National Academy of Sciences, 17, 57, 126
natural products, 39, 40, 41, 44, 46, 54, 55, 58, 59
nitrogen, 33, 40, 43, 91, 137
nonspecific stretegies, 1
nucleic acid, 12, 28, 69
nutrient(s), 7, 23, 27, 41, 44, 116, 117, 118, 119, 120, 122, 127, 133
nutrition, 116, 125, 140

O

organic solvents, 84, 101, 110
organism, 122, 132, 134, 137
osmolality, 72, 77, 140
osmotic pressure, 71, 133, 135, 136
osmotic stress, 70, 75, 133, 136, 137, 141
oxidation, 29, 32, 33, 84, 94, 95, 97
oxidative stress, 20, 27, 29, 30, 32, 40, 43, 48
oxygen, 6, 13, 26, 30, 33, 36, 37, 40, 43, 84, 106, 117, 125

P

pathogenesis, 4, 34, 70
pathogens, 17, 30, 42, 43, 121
pathway(s), 2, 27, 28, 30, 33, 34, 35, 38, 52, 74, 90, 101, 103, 120
peptide(s), 17, 44, 45

pH, 6, 7, 12, 13, 14, 18, 19, 20, 21, 22, 26, 40, 43, 62, 69, 73, 84, 86, 94, 99, 105, 110, 112, 117, 132, 135
pharmaceutical, viii, 40, 44, 86
phenol, 84, 99, 101, 108
phenotype(s), 2, 66, 69, 94, 123
phospholipids, 10, 64, 65
phosphorylation, 12, 20, 27, 28, 78
physiology, vii, 5, 9, 12, 16, 22, 36, 41, 55, 61, 62, 64, 65, 68, 69, 71, 72, 73, 74, 116, 117, 118, 125, 138, 141
plants, 40, 84, 97, 142
plasma membrane, 61, 62, 68, 73, 80, 135
polymerase, 46, 48, 49, 58
polymers, 66, 71, 101
polysaccharide(s), 53, 120, 121, 129
potassium, 17, 134, 135
production of antibiotics, 40, 41, 43, 58
prokaryotes, 46, 63, 78, 141
promoter, 46, 48, 49, 50, 54
protein engineering, 84, 95, 112
protein structure, 69, 89, 90
protein synthesis, 10, 76, 135
proteins, 9, 10, 11, 13, 27, 28, 29, 30, 32, 33, 37, 38, 51, 62, 63, 64, 67, 68, 70, 72, 79, 84, 86, 89, 93, 97, 108, 111, 120, 121, 134, 135, 136, 137, 142
proteome, 63, 122, 141
Pseudomonas aeruginosa, 10, 21, 67, 73, 77, 78, 79, 80
pumps, 10, 67, 78, 79

Q

QM-MM, 84, 103

R

radiation, 7, 62, 79
reactions, 6, 12, 20, 69, 97, 100
receptor(s), vii, 28, 31, 73, 124, 128
recognition, 22, 46, 48
relevance, vii, viii, 2, 5, 13, 15, 25, 65, 71, 75, 116, 120, 122, 131
replication, 41, 51, 135
residue(s), 64, 86, 92, 95, 97, 98, 99, 100, 103, 108, 121, 134
resistance, 4, 8, 9, 10, 11, 16, 17, 21, 22, 23, 27, 31, 34, 42, 49, 65, 68, 73, 76, 78, 79, 95, 102, 107
response, vii, 2, 4, 6, 7, 8, 15, 26, 27, 31, 32, 33, 36, 48, 54, 62, 63, 70, 72, 73, 118, 123, 127
ribosome(s), 12, 21, 41, 50, 51, 52, 69
RNA, 22, 41, 46, 48, 49, 58, 75, 76, 77, 122
RNA polymerase holoenzyme, 46

S

salinity, 5, 62, 133, 137, 138
Salmonella, 19, 69, 74, 76, 77, 80, 137
salt concentration, 2, 70, 84, 132, 133, 134, 136, 139, 141
salts, 132, 133, 138
scarcity, 41, 46, 137
science, 76, 128, 129
secondary metabolism, 30, 32, 42, 47, 49, 50, 52, 120
sensing, 7, 32, 35, 43, 56, 64, 66, 70, 73, 75, 123, 125, 127
sensor, 32, 33, 75
sequencing, 41, 44, 128
sigma factor(s), 40, 41, 43, 46, 47, 48, 49, 51, 56, 57, 58
signal transduction, 27, 40, 47, 49, 58
signalling, 27, 28, 29, 30, 35, 36, 38
signals, 15, 28, 30, 40, 43, 48, 54, 117
simulation, 94, 101, 103, 106, 108, 109, 110, 111
simulations, 84, 86, 100, 102, 110
smBGCs, 42, 44, 46, 55

species, vii, viii, 3, 27, 33, 34, 39, 40, 42, 43, 44, 47, 48, 55, 59, 68, 70, 90, 115, 118, 120, 121, 123, 124, 133
stability, 3, 16, 65, 96, 101, 102, 103
stabilization, 12, 69, 85, 99
state(s), 19, 26, 36, 64, 84, 87, 95, 97, 112, 116, 117
streptomyces, v, 39, 40, 41, 42, 43, 44, 45, 46, 47, 48, 49, 52, 53, 54, 55, 56, 57, 58, 59, 68, 89, 97, 104, 105, 109
Streptomyces avermitilis, 53, 56, 57
stress, v, 1, 2, 6, 7, 15, 18, 19, 20, 23, 27, 28, 29, 30, 31, 32, 33, 34, 36, 37, 38, 39, 40, 43, 48, 54, 56, 58, 61, 62, 65, 67, 70, 72, 73, 74, 75, 77, 79, 80, 81, 105, 133, 134, 136, 137, 138, 140, 141
stress response, 15, 28, 29, 31, 34, 36, 56, 58, 72, 77, 140
structure, 7, 15, 20, 22, 57, 61, 62, 70, 71, 73, 84, 87, 88, 89, 90, 91, 92, 93, 94, 97, 99, 101, 102, 103, 105, 107, 111, 112, 117, 121
substrate(s), 3, 7, 16, 26, 77, 85, 89, 94, 97, 98, 109, 121, 123, 131
sucrose, 120, 136, 137
Sun, 41, 43, 44, 46, 47, 48, 49, 50, 58, 59
survival, 14, 16, 26, 35, 48, 74, 135
synthesis, 13, 29, 35, 54, 55, 75, 76, 77, 78, 85, 117, 135, 140

T

target, 9, 11, 20, 21, 49, 53
techniques, viii, 9, 25, 44, 93, 94, 97, 111, 120, 125
temperature, 4, 5, 27, 40, 43, 62, 85, 95, 105, 132
transcription, 10, 16, 27, 29, 30, 33, 41, 42, 46, 47, 48, 51, 53, 54, 58, 59, 69
transcription factors, 16, 27, 69
transcriptional factors, 40, 41, 42
transduction, vii, 27, 40, 47, 49, 58, 73, 124
translation, 40, 41, 43, 49, 50, 51
transport, vii, 2, 10, 12, 14, 18, 19, 62, 66, 68, 69, 73, 76, 77, 80, 138, 142

V

variables, 1, 5, 6, 12, 15, 62
viruses, 5, 19, 120
viscosity, 63, 64, 80

W

water, 5, 17, 27, 66, 70, 72, 74, 84, 101, 103, 132, 133, 134, 137, 138, 141

Y

yield, 31, 32, 34